30 Day Paleo Challenge

30 Day
Paleo
Challenge

Unlock Your Weight Loss Secret with the Paleo 30 Day Challenge; Paleo Cookbook with 30 Day Meal Plan and 100 Paleo Recipes

Irene Kadison

Want MORE healthy recipes for FREE?

Double down on healthy living with a full week of fresh, healthy salad recipes. A new salad for every day of the week!

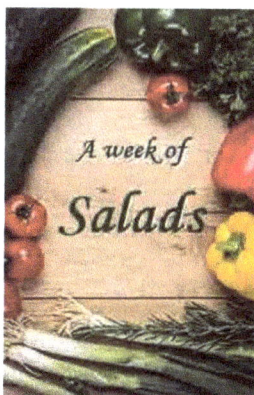

Grab this bonus recipe ebook *free* as our gift to you:

http://salad7.hotbooks.org

Want MORE full length cookbooks for FREE?

We invite you to sign up for free review copies of future books!

Learn more and get brand new cookbooks for **free**:

http://club.hotbooks.org

Want to kick it up a notch?

Double down on healthy living with a full week of fresh, healthy fruit and vegetable juice recipes. A new juice for every day of the week!

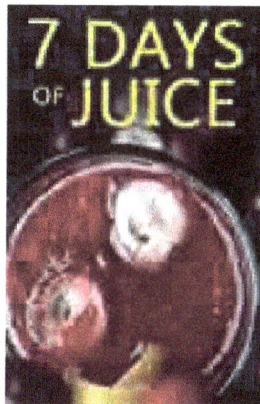

Grab this bonus recipe ebook *free* as our gift to you:

http://juice7.hotbooks.org

Contents

INTRODUCTION

To The Paleo Lifestyle

Introduction

The modern era has brought humanity a lot of good things, like the internet and improved healthcare, but food and nutrition is one area where we've really taken a steep dive. It seems next to impossible to get food that isn't packed with artificial ingredients or stripped of its natural benefits. What is a person who wants to get healthy to do?

The Paleo lifestyle turns back the clock and looks at what humans were eating 2 million years ago. The idea behind "going Paleo" is that humans were healthier back then. Certain food groups like grain and dairy had yet to be introduced, and once they were, the trouble started. By eliminating certain foods, a modern person can get back some of their health and enjoy a host of benefits including improved weight loss, mental clarity, better sleep, and so on.

This book explores how the Paleo lifestyle became so popular, and what separates it from other "fad" diets like Whole30 or Atkins. You'll learn what you can and can't eat, and what is on the "maybe" list. The Paleo lifestyle is flexible and can be customized, which is good news for people who want to maintain the eating lifestyle for the long-term. You'll learn how to read labels, how to grocery shop, and what kitchen tools can make home-cooking fun and convenient.

Going Paleo is a big transition, especially for those who have been relying on gluten-heavy, packaged foods. The first weeks can bring on what's known as the "carb flu." This book will teach you how to treat the symptoms with organic bone broth, slow-burning carbs, and so on. Once you've passed through the carb flu, you'll feel cleansed and energized.

The Paleo lifestyle is not without its problems, which this book examines closely. If you decide the Paleo lifestyle is in fact a good option for you, take a good look at the general tips for success. These are your guiding principles for as long as you're eating Paleo, and help ensure that changing your diet doesn't disrupt your life. It should transform it, so you feel energetic, healthier, and ready to take on the world.

HISTORY OF THE PALEO LIFESTYLE

The idea of eating like humans in the distant past has been around since the 1970's and 1980's, but it wasn't until recently that the diet really took off. The "Paleo lifestyle" as we know it came into being in 2002, when a doctor began investigating diets that could prevent the chronic health conditions associated with processed food. Dr. Cordain went back into history - far back - and learned that our Paleolithic ancestors were in much better shape than modern humans. They were on their feet all day, athletic, and didn't appear to have many of the diseases that afflict us today. What was their secret?

Dr. Cordain looked at the Paleolithic diet and found that it was very high in fiber and protein, but included very few grains and little to no sugar. Cavemen lived on meat and vegetables, and whatever fruit, nuts, and seeds they could find. As nomads and hunters, they were extremely active, and spent their days hiking, running, and climbing. Their bodies burned fat and protein for energy. They went to bed when the sun set and got up when it rose again.

As soon as humans settled into more organized groups, and began to farm, the world changed forever. While still active, their diet began to include the grains they grew and dairy from domesticated animals. Our bodies began to change and evolve, too, but according to Dr. Cordain, it wasn't enough.

Cut to today, and our bodies haven't adjusted enough to handle the high volume of grains and dairy that we eat. We've made things much worse by also consuming processed foods and tons of sugar. We've developed technology that lets us sit around all day, so all that junk turns to stored fat. The results? Obesity, increased vulnerability to disease, lower energy, mental health problems, digestive distress, and more.

WHAT DOES IT MEAN TO GO PALEO?

Dr. Cordain promoted a diet that was free of dairy, grains, processed foods, and any type of vegetable or fruit that wasn't available to Paleolithic humans. He broke it down into a 55-15% split, with the higher percentage devoted to seafood and lean meat, and the remaining three 15% chunks coming from fruit, vegetables, and nuts and seeds. While some people like the percentage breakdown, it isn't required to be on the diet.

Calorie-counting is also not recommended on a true Paleo lifestyle, so as a general rule, remember to eat a lot of fiber and fat, a good amount of protein, and a moderate amount of carbs, which will pretty much all come from vegetables.

It's important to note that "going Paleo" has bloomed into more than changing your eating habits. Paleo has inspired an entire way of thinking that's based on simplicity and breaking free from technology. There are magazines, conferences, and retreats all devoted to the Paleo lifestyle, which includes homemade shampoo and beauty products, "barefoot" shoes and clothing, and more. Exercising is all about doing what Paleolithics did naturally - lifting things, running, and climbing. No fancy gym equipment required.

WHAT SEPARATES PALEO FROM OTHER "FAD" DIETS?

There are lots of diets in the headlines that have similar buzzwords, so it's hard to know what the difference is between them. Some let you eat dairy, while others cut it out completely. One diet might want you to count calories, while another is more concerned about macros. Where does Paleo line up?

The "clean eating" diet

People who "eat clean" aren't following any particular diet and there aren't any rules beyond sticking to organic food. That means grass-fed meats, whole grains, and so on. No food groups are eliminated, and there aren't any guidelines about counting calories or macronutrients.

Whole Food 30 Day Challenge

This is a very restrictive diet designed for just 30 days. You can't eat any sugar, dairy, grains, alcohol, any baked goods (including homemade with Whole Food approved ingredients), legumes, or anything with artificial or processed ingredients. The purpose of the diet is to reset the body. Many people use it lose a lot of weight very quickly or identify food allergies.

Atkins diet

The Atkins diet has had many evolutions, but the core concept is that it's designed for weight loss and has you eat very little carbs. You can eat as many proteins and fats as you want. It's been harshly criticized over the years, and the fact that the creator died of a heart attack doesn't help its credibility.

The Ketogenic diet

The Keto diet overlaps in a few ways with the Paleo lifestyle, most notably that both eliminate grains, but the Keto diet is all about getting the body into "ketosis." Ketosis is when the body depends on fat instead of carbs for energy, so full-fat dairy is included and encouraged in the Keto diet. The Keto diet also has you balance three macronutrients - fat, carbs, and protein - in a specific way, while the Paleo lifestyle does not.

The Bulletproof diet

This diet is all about fat (losing it) and muscle (building it). It's supposed to raise energy levels, improve mental clarity, and reduce inflammation. The creators divide foods into "zones," with green-zone foods being the best, and red-zone ones the food to avoid. Green-zone foods include grass-fed butter, lots of vegetables and animal protein, and coconut oil. The most famous/infamous part of the Bulletproof diet is "bulletproof coffee," which is coffee with butter and "brain octane" or MCT oil.

Paleo

While the other diets are relatively new, supporters of Paleo say it works because it's based on thousands of years of evidence. The point of the diet isn't to lose weight, though a lot of people use it that way. The ultimate goal is to transform the body back into its natural state of strength and health.

WHAT YOU EAT ON PALEO

On Paleo, the bulk of what you're eating will be meat and vegetables. It's pretty simple, and to feel full, you'll find that you will be eating a lot of vegetables. In terms of nutrition, always remember protein, fiber, and fat. If you're feeling hungry during the way, even with hearty meals, eat something fatty like an avocado or nuts. You can also add more fat to your meals. Here's a list of what you are allowed to eat:

Meat

To get meat in its purest form, without additives, get grass-fed, pasture-raised, and organic meat whenever possible. Don't be afraid about fatty cuts, since animal fat is awesome on Paleo.

Includes beef, bacon, lamb, goat, chicken, turkey, veal, pheasant, rabbit, wild boar

Fish/seafood

Like with meat, you want to get the best quality whenever you can, which means getting wild-caught fish as often as possible.

Includes clams, salmon, shrimp, crab, oysters, mackerel, scallops, sardines, tilapia, trout, tuna, walleye

Eggs (organic)

Brown or white

Vegetables

Pretty much all vegetables - fresh or frozen - are allowed on Paleo. Eat starchy vegetables like squash and sweet potatoes in moderation, because they don't pack in as much nutrition.

Includes avocado, broccoli, artichoke, acorn squash, asparagus, Brussels sprouts, cabbage, carrots, cauliflower, cucumber, eggplant, kale, lettuce, onions, spinach, pumpkin, celery, bell peppers, sweet potatoes, parsnip, zucchini

Fruits

A lot of fruit is very high in sugar, so you should eat in moderation. Some people eliminate fruit altogether if they are trying to lose weight. The best fruits are berries, because they are rich in antioxidants and have relatively low sugar.

Includes apples, blueberries, bananas, blackberries, cranberries, figs, grapes, lemons, limes, mango, oranges, peaches, pineapple, raspberries, strawberries, watermelon

Nuts/seeds

Nuts and seeds are good Paleo foods because of their high fat content. However, if you have digestive problems, consider eliminating nuts from your diet. You don't need to eat nuts and seeds on Paleo, but the fats can help satisfy you more if you're feeling hungry.

Includes almonds, cashews, walnuts, pumpkin seeds, sunflower seeds, pecans, pine nuts, macadamia nuts (a good nut choice if you want to keep just one kind), hazelnuts

Oils/fats

Healthy fats are essential on Paleo. Some say you should really only cook with ghee (clarified butter) or animal fat. Olive oil and coconut oil are allowed as substitutes for salad dressings, and can be poured on food after you've cooked it. However, other Paleo foodie experts say it's okay to cook olive oil, so it's up to you.

There's also some discussion on MCT oil, which stands for medium-chain triglyceride oil. This oil is derived from coconut oil and claims to have a higher concentration of fatty acids. Unless you're trying to add ketosis into your Paleo lifestyle, it isn't really necessary.

Includes ghee, avocado oil, olive oil, coconut oil, macadamia nut oil, high-quality duck fat, beef tallow

Beverages

Water, herbal tea, fruit juice (that you juice yourself, but in extreme moderation - can be used as sweetener)

WHAT YOU CAN'T EAT

Most of what you can't eat makes sense, like anything processed or packed with artificial ingredients. However, you also can't eat traditionally "healthy" foods like whole grains and dairy. We'll get into the "why" in more depth in the section on the Paleo lifestyle and health, but just know that there are nutritional reasons that make grains and dairy undesirable.

Dairy

Includes butter, cheese, cottage cheese, milk, ice cream, yogurt, powdered milk

Grains/legumes

All beans, peanuts, tofu, soy, lentils, peas, bread, chickpeas, cereal, miso, oatmeal, quinoa, rice, pasta

Candy/snack foods

Chips, candy bars, chocolate chips, granola bars, energy bars, crackers

Processed deli meats

Spam, deli ham, deli turkey, roast beef

Oils

All vegetable oils, all bottled salad dressings

All sweeteners

Sugar, artificial sweeteners, natural sweeteners (stevia, erythritol), vanilla extract, other flavor extracts

Drinks

Energy drinks, sodas, alcohol

WHAT YOU CAN (MAYBE) EAT

Members of the Paleo community debate a lot about certain foods. Much of the discussion has to do with newer research on the Paleolithic era, that maybe the original guidelines missed some things, like raw honey, which many experts say cavemen would have taken from hives they found. The same goes for goat's milk and butter, that perhaps cavemen were using animals for these purposes earlier than originally believed. Here's a list of those foods:

- Homemade treats (pancakes, cookies, candy) made with Paleo-friendly ingredients
- Vinegar (white wine, balsamic, apple cider, red wine)
- Grass-fed butter
- Buttermilk
- Paleo-friendly bread
- Bittersweet chocolate
- Green beans (technically a legume)
- Beansprouts
- Potato flour/potato starch
- Sesame seed oil
- Almond milk
- Goat's milk
- Whey protein
- Raw honey
- Maple syrup
- Fruit jam
- 100%, no-sugar added fruit juice
- Whisky
- Black coffee

Hardcore Paleo purists are more concerned about making "treats" with Paleo ingredients than whether or not cavemen ate honey or maple syrup. They like to say that just because you can make something with Paleo ingredients, it doesn't actually make it Paleo. It's not as if cavemen were making pancakes with the foods they were eating and pouring on raw maple syrup. However, it's up to you with what you're okay with, and how far you want to take the Paleo mindset.

GROCERY SHOPPING

When you're starting on Paleo, and you go to the store, it can be overwhelming. Where do you start? What sections are going to have the most Paleo-friendly items? Do you need to go to a special store? There are lots of tips and tricks that can make the trip easier, and each time you go back, it gets less time-consuming:

Always plan ahead

This is just good advice for any shopper. Make a meal plan and write a list of all the ingredients you'll need before hitting the store. This way, you won't forget anything and need to come back, and you won't be as tempted to throw random things in your cart.

Double-check your kitchen

Before you head off to the store, take your list around your kitchen, fridge, and pantry to make sure you don't have any of the items already. This saves money, and prevents food waste, which is a major issue nowadays. If you're invested in the Paleo lifestyle, you probably want to use everything you can, and the idea of food waste is shameful.

Paleo stuff is usually around the edges of the store

As a general rule, all of the junk food is in the center of stores, where people tend to wander. If you stick around the edges, you're going to find the produce, frozen foods, and organic section.

Buy in bulk

When you buy in bulk, you can save a lot of money in the long-term. Make sure you're getting items you use a lot, that you can freeze (like meat), or that will last a long time. You don't want to buy something in bulk and then never use it.

Buy seasonally

The best way to get good deals on produce is to buy what's in season. Look online for information on what's in season, and then decide on recipes that way. When you go to the store or farmer's market (my favorite place to get produce), get what's in season. It will be super fresh and affordable.

Buy online

Another way to get the best deals on stuff like grass-fed beef is to shop online. There are lots of good sources that have free shipping and are priced lower per pound than grocery stores. It's a great way to buy in bulk, so you can freeze and then thaw as you eat. Amazon Pantry is also a good resource, and has some good brands you can't find elsewhere.

Get frozen vegetables/fruit

It's pretty easy to find Paleo-friendly organic frozen vegetables and fruit just about anywhere. They aren't seasoned, so you don't have to worry about dubious ingredients, and they're easy to prepare.

"Weird" cuts of meat are cheaper

Organ meats and other "odd" cuts of meat are usually cheaper than the standard breast, tenderloin, etc., because most people don't go for them. Learn to appreciate liver, tongue, heart, and more to save money and get lots of nutritional value. Bone-in meat is also usually more affordable than boneless.

Organic stores will have more options

While you can shop at chain stores and find Paleo foods, there won't be a large selection. To find the best and the most, you'll probably have to start going to a higher-end store like Whole Foods. This can get pricy, so you want to be on the lookout for coupons and sales. Trader Joe's is a more affordable choice that is much loved for the quality. Some other good stores include: Aldi, Wegman's, Sprouts, and Costco.

Get good at couponing

Couponing is going to be one of the best ways you can save money, especially when you want to buy in bulk. Break out the scissors and start clipping the old-fashioned way. There are coupons online, as well, but the vast majority of deals are still found stuck in mailboxes and in newspapers.

Write down the Paleo-friendly foods you find

When you're starting on the Paleo lifestyle, write down everything you find that's approved, especially if it's in a box or bag. You'll remember that organic lemons are Paleo, but what was that specific brand of almond flour you liked? Compiling a list right at the start will save you time and energy.

READING LABELS

An important part of going Paleo will involve reading labels, so you can figure out what packaged or bagged foods are going to be allowed on your diet. While the majority of your food will simply be whole foods (like beef, veggies, and nuts), you will still need to use food that comes in jars and boxes. Here are some tips on labels:

If you don't know what it is, you probably shouldn't have it

When you're looking at a label and come across a word you don't recognize as food, it's almost certainly not allowed on Paleo. Anything with numbers is a dead giveaway that the ingredient does not appear in nature.

If it has sugar, don't buy it

Added sugar is a huge issue with packaged foods, so look and see where it falls in the ingredient list. If it's first or second, it's a definite no, and if you're being really strict about being Paleo, if it appears anywhere on the list, it's still a no.

Check the source of carbs

Look at the nutrition label and see how many carbs are in a food item. Carbs aren't necessarily bad on Paleo, but where are they coming from? Are they coming from a fruit, vegetable, or nut source? If not, it's most likely coming from sugar.

If it has additives and preservatives, skip it

Food additives are usually meant to preserve or add flavors or coloring, which is definitely not Paleo. Sulfates, MSG, caramel coloring, and so on all mean that a food is not Paleo.

The shorter the ingredient list, the better

When you're going Paleo, the shorter an ingredient list is, the better. A short list tends to mean the food relies on only whole foods like fruit, nuts, and so on, while longer lists mean the food is packed with artificial preservatives and fillers.

CODE WORDS FOR WHEAT, SOY, SUGAR, AND CORN

Stuff like gluten and sugar aren't always labeled clearly on food, so it's important to know all the different names companies will use. Some make sense, like corn syrup, but others are less obvious. Here's a handy list you can use when you're shopping. If any package you pick up has any of these ingredients, it's not Paleo-friendly:

Wheat/gluten

Dextrin, caramel color, artificial flavoring, malt, hydrolyzed wheat protein, hydrolyzed wheat starch, hydrolyzed plant protein (HPP), hydrolyzed vegetable protein (HVP), maltodextrin, vegetable starch, natural flavoring,

Soy

Artificial flavoring, miso, natural flavoring, HPP, HVP, stabilizer, tamari, tempeh, tofu, vegetable broth, textured vegetable protein (TVP), vegetable starch, vegetable gum

Sugar

Cane crystals, barley malt syrup, dextrin, dextrose, disaccharide, galactose, lactose, maltodextrin, maltose, monosaccharide, polysaccharide, ribose, saccharose, sorghum, sucrose, xylose

Corn

Artificial flavoring, dextrin, dextrose, food starch, mazena, modified gum starch, sorbitol, MSG, xanthan gum, xylitol

PALEO-FRIENDLY BRANDS

To save you on some research and effort, here's a list of brands we know are Paleo-friendly. Because they're high-quality, they cost more than the average packaged food, but they're totally worth it.

- Artisana - coconut butter, almond butter
- RXBAR - protein bars offering 12 grams of protein w/ very simple ingredients
- Epic Bars - protein bars made with meat and dried fruit
- Larabar - banana bread, apple pie, cashew cookie, and pecan pie are Paleo
- Jilz Crackers - grain-free
- Seasnax - seaweed snacks made w/ olive oil
- Zupa Noma - drinking soup made with vegetables in a convenient bottle
- The New Primal - meat and jerky snacks, all grass-fed
- Primal Kitchen - salad dressings
- Saucy Lips - all Paleo marinades
- Thai Kitchen - coconut milk
- Kettle and Fire - chicken, beef, and mushroom chicken bone broths
- Bob's Red Mill - almond/coconut flour + arrowroot powder
- Nutpods - half almond milk, half coconut milk creamers
- New Barn - unsweetened almond milk

- Califia Farms - unsweetened almond milk
- Trader Joe's - lite coconut milk/organic coconut milk
- Wild Planet - all natural and organic canned meat and fish
- Pederson's Natural Farms - all natural meats like bacon and breakfast sausage
- Tcho 99 - dark chocolate
- Steve's Original - granola, jerky, bars, etc
- Paleonola - granola cereal substitute

Watch out for Designs for Health "PaleoBar," which has sugar and gluten, and Caveman Foods, whose Caveman Crunch bars have tons of added sugar. Even if something is labeled "Paleo," read the label to see for yourself. There are lots of apps that can help you learn if something is Paleo-friendly or not; we'll list those a bit later on.

STOCKING A PALEO KITCHEN

What you get specifically depends on what your list looks like (you don't have to get chuck roast every time, but you should always have ground beef and another cut every time) and what's in season, but in general, this is what a typical Paleo list should look like:

Produce (organic whenever possible)

- Tomatoes
- Dark leafy green (like kale or spinach)
- Carrots
- Cauliflower
- Onions
- Garlic
- Bell peppers
- Zucchini
- Cucumbers
- Celery
- Berries
- Lemons
- Limes

Meat (grass-fed/pasture-raised)

- Ground beef
- Chuck roast/steak
- Bacon
- Chicken breast or thighs
- Fish (fresh + frozen)
- Tuna (canned)

Dairy

- Organic eggs

Oils

- Coconut oil
- Extra virgin olive oil
- Grass-fed butter (maybe) Or ghee

Nuts/seeds

- Macadamia nuts
- Raw almonds or cashews
- Sunflower seeds
- Ground flaxseed

Baking supplies

- Coconut flour
- Almond flour
- Arrowroot powder
- Baking powder
- Baking soda
- Raw honey (if you're allowing it)
- Dark chocolate (if you're allowing it)
- Almond butter (if you're allowing it)

Canned/cooking supplies

- Full-fat coconut milk
- Unsweetened shredded coconut
- Organic diced tomatoes
- Balsamic vinegar
- Apple cider vinegar

Spices/herbs

- Salt
- Black pepper
- Dried oregano
- Dried thyme
- Dried sage
- Garlic powder
- Onion powder
- Cumin
- Turmeric
- Chipotle powder
- Chili powder
- Ground clove
- Ground nutmeg
- Ground ginger
- Ground cinnamon
- Smoked paprika
- Dried dill
- Crushed red pepper flakes

Beverages

- Herbal tea
- Unsweetened coconut water
- 100% fruit juice (for use as a sweetener)

Reasons why the Paleo lifestyle is a good choice

Improving one's health was the driving force behind the Paleo lifestyle's creation. Processed and artificial foods cause countless health problems, and a diet that cuts out problem food while embracing what's good is going to have benefits. There are several results that people focus on:

Healthier heart

A two-week study with 36 men and women with an average age of 54 examined Paleo and a "healthy reference" diet with whole grains and low-fat dairy. The researchers found that the Paleo lifestyle lowered blood pressure, raised HDL cholesterol, and lowered the bad cholesterol. This makes sense, considering the Paleo lifestyle is rich in food that proven health benefits, such as spinach, avocado, salmon, extra virgin olive oil, and blueberries.

Improved weight loss

When people change their diets, they often want to lose weight. There have been quite a few studies on whether a Paleo lifestyle can help, and there's been good results. One study compared a diet tailored to diabetics to a Paleo lifestyle. There were 13 volunteers (each with type 2 diabetes) who were on the diet for three months. The volunteers on Paleo lost 6.6 more pounds and 4 centimeters off their waists than those on the diabetes diet. In another study, 5 men and 9 women stuck to a Paleo lifestyle for 3 weeks, and on average, they lost about 5 pounds.

The focus on healthy Omega 3 and Omega 6 helps burn off stored body fat, while the Paleo mindset encourages exercising, especially outdoors. This allows the body to get a good amount of Vitamin D (assuming you can exercise in the sun) and speeds up the body's metabolic processes in general.

Reduced inflammation

Chronic inflammation is responsible for a wide range of health problems, from joint paint to stomach problems to an increased risk of cancer. While acute inflammation is a natural response to a wound or other injury, chronic inflammation is what happens when that response goes on longer than it's actually needed. White blood cells, which are intended to heal damage and attack foreign substances, start attacking tissues, cells, and even internal organs. Inflammation can also occur when you aren't getting enough nutrition and/or your body can't absorb it because of damage from toxic foods, like excess sugar.

A Paleo lifestyle cuts out a lot of the foods responsible for that damage and inflammatory response, like gluten and dairy, and foods high in saturated and trans fats. Eating Paleo means eating lots of foods like dark leafy greens, tomatoes, fish, and nuts, which reduce inflammation.

Better sleep

Sleep is essential to a healthy body and mind. When the amount of carbohydrates, especially the kinds of processed food and grain, the body's blood sugar levels become more stabilized. This allows a person to fall asleep faster, and stay asleep. Studies have also shown that a Paleo lifestyle can help with diseases that result in insomnia, such as diabetes.

Going beyond just what one eats, the Paleo lifestyle encourages healthier sleep patterns. Our cavemen ancestors went to bed when the sun went down and got up when it rose again. Our bodies were meant to follow these patterns, so when we modify our schedules, we can enjoy better and deeper sleep.

Improved mental clarity

The Paleo lifestyle is very good for the brain, thanks to Omega 3 fatty acids. The fact that the Paleo lifestyle also helps regulate blood sugar is important, because many medical experts say that blood-sugar regulation is the most important key to preserving proper brain function. It makes sense that what we eat affects our brains - after all, food is fuel. When you cut out processed food and other foods that the body isn't great at using, problems like anxiety, "brain fog," and more will improve.

More satisfaction after eating

The last benefit of going Paleo is that your meals will satisfy you and keep you full longer. There are a couple of reasons for this, including the fact that a diet full of vegetables has a lot of fiber, which is known to give you that full feeling. The slow-burning carbs also help. While fast-burning carbs in processed foods burn off quickly, leaving you feeling hungry again, slow-burning carbs stay in your system longer.

CRITICISMS OF THE PALEO LIFESTYLE

As with every diet, especially ones that become very popular very quickly, there are problems that nutritionists and experts have pointed out. There's been questions about the science behind what Paleo means, as well as issues about just how practical the diet is, or if it's even healthy. Here are the most common concerns that people raise:

It's impossible to actually eat truly Paleo, because of how our food has evolved.

Detractors of Paleo like to point out that "going Paleo" is technically impossible, because the meat and plants we eat today are a result of thousands of years of evolution. You cannot get a lemon that looks like the lemon a cavemen would have eaten, because humans have always bred plants to grow the best and tastiest fruits, and bred animals the same way, too. This would mean that people on Paleo aren't going to be getting any of the benefits that cavemen would have, but there are still countless proven benefits to eating certain foods, especially organic and grass-fed, so this isn't an especially convincing criticism.

It's a restrictive diet, which is hard to maintain for the long-term.

The Paleo lifestyle cuts out entire food groups - dairy and grains - as well as other foods, which can make sticking to the diet very difficult in the long-term. It goes beyond just not drinking milk or eating traditional bread, because dairy and grains are ingredients in tons of food, including ones that your friends and family will prepare. It can get exhausting to always be asking about ingredients, bringing your own food everywhere, and so on. This is why the Paleo 80/20 is so appealing to people. They can be strictly Paleo when they're in charge of their food, and then when they're eating at someone's house, they can eat whatever is put on the table.

Buying organic, grass-fed, and wild-caught food gets expensive.

A big part of the Paleo lifestyle is the quality of foods that you get. That means organic veggies and fruit, grass-fed meats, and wild-caught seafood. That can get really expensive, and not everyone has a budget that can accommodate paying $3.00 for a pound of organic zucchinis every week or $2.50 more per pound on grass-fed beef. The solution is to be smart about what's in season, going to places like farmer's markets and farmers for produce and fruit, or shopping online for grass-fed meats.

More research into the Paleolithic era has cast doubt on whether humans were healthier than today.

The foundational idea behind the Paleo lifestyle is that cavemen were healthier than humans today. However, more research has shown that the average lifespan for Paleolithic humans was only about 18. There's also been a study that showed that 47 out of 137 mummies from the Middle Age showed signs of heart disease, specifically hardened arteries, which led researchers to conclude that heart disease might be a normal part of aging and not connected to any particular type of diet.

However, the short lifespan of cavemen can be attributed to dying in childbirth, getting killed by other humans or animals, and succumbing to infections and other now easily-cured ailments. While "healthier" maybe isn't the right word, there are definitely certain health problems humans have today that didn't exist for cavemen.

Lots of healthy nutrients, like Vitamin D and calcium, are excluded when you don't eat dairy, whole-grains, and beans.

One of the biggest concerns experts have about the Paleo lifestyle is that cutting out entire food groups increases a person's risk of mineral and vitamin deficiency. Vitamin D and calcium are two of the big ones. When you don't consume dairy, you miss out on both Vitamin D and calcium. That can make you vulnerable to dementia, heart attacks, osteoporosis, and more.

However, you can eat Vitamin D and calcium from other sources. Dark leafy greens are full of calcium, as are almonds and seafood. Vitamin D can be found in fatty, cold-water fish like salmon, and you can Vitamin D from the sun. If all else fails, you can also take high-quality supplements.

KITCHEN MUST-HAVES

In theory, you don't need a lot of kitchen equipment to eat Paleo, but it certainly makes your life easier. Whether you're slicing, chopping, spiralizer, or steaming, having the right tool makes cooking fun and convenient. Here's what you should consider getting:

A little food processor

A small, affordable food processor like a Ninja Chop is a fantastic tool that has multiple purposes. You can make single servings of green smoothies and soups, and prepare ingredients like nuts to replace breading or chop up veggies really quickly. If you want to be able to make larger portions, you'll need to spend more on a real blender, but if you don't think that's necessary, a small processor is the next best thing.

Cast-iron pan

These last forever, so they can be a bit pricy, but they don't break down they cook meats and veggies more evenly than other types of pans. Cast-iron is safe in the oven, too. Be sure you read about how to properly season cast-iron and clean them, so it lasts as long as possible.

Digital pressure cooker

If you've never used a digital pressure cooker, you are majorly missing out. While a bit pricier than your average slow cooker, pressure cookers are up to three times as fast and have been proven to be the healthiest cooking method. There are stovetop or digital pressure cookers, but for convenience's sake, digital is really the way to go. You can literally cook anything in them, and many have the option to switch to slow-cooker mode, so it's really two devices in one.

Good knives

You'll be doing a lot of chopping and slicing and dicing when you're on Paleo and making everything at home, so having good knives is crucial. Knives were meant to be an investment, so don't get the cheapest kind you can find. Whenever something is coming into contact with your food, you want to make sure it's high-quality. You'll probably have to choose between high carbon or stainless steel. High carbon knives are easier to sharpen, but require more care, while stainless steel knives are harder to sharpen, but are more convenient for the average person.

Spiralizer (or mandolin)

Spiralizer and mandolins are similar, but the main difference is the shape it makes from the vegetable. Spiralizers create curly strands or ribbons, like pasta, while mandolins just slice. If you want to make a lot of veggie noodles, the spiralizer is probably the better choice.

Durable storage containers

With nearly all your meals being homemade, you'll be taking them to places like work most days. That means you'll need good storage containers that will last and won't taint your food with chemicals. Some people like glass, because they can put it in the dishwasher, while silicone is less fragile. Whatever you get, make sure it's at least microwave and freezer-safe.

PALEO APPS

In our modern age, one of the coolest innovations has been apps. There are many devoted to the Paleo lifestyle that make it easier to educate yourself and connect with others while on the go. Here are some to consider downloading on your smartphone:

Paleo.io

You probably don't want to spend every grocery trip scanning labels, so an easy way to check if a food is Paleo is to get the "Paleo.io" app for IOS devices. You type in the food and the app should tell you if it's okay. The app has other information like recipes and eating guides. "Paleo Food List" is another database with over 6,000 foods.

Primal Paleo

This $1.99 app is packed with features and is considered the "most complete Paleo and lifestyle guide." It has a list of all the foods allowed on Paleo, workouts, and recipes.

Paleoviz

For those who want to track their health goal progress, this $1 app is a great tool. You record your meals, and the app grades it based on how Paleo it is, complete with a comprehensive scorecard after you've tracked a few meals.

Paleo Leap

With over 900 recipes, this app has gorgeous photos and meal plans. You can also search for recipe specifics, like AIP or slow cooker.

Nom Nom Paleo

No Paleo app list would be complete without this award-winning app that has 145 recipes with thousands of photos that walk you through each step of the preparation process. The app also lets you write grocery lists.

PALEO MEAL DELIVERY

If you've been interested in getting a meal delivery service, there are ones entirely devoted to the Paleo lifestyle! Meal delivery kits in general tend to be a bit pricey, but if you've been using one and want to switch to Paleo, or you decide a service fits your budget, here are ones to check out:

Trifecta Nutrition

This service partners with Dr. Loren Cordain, the creator of Paleo, so you can know for sure it's 100% Paleo. You can order food in bulk if you don't want meals, and Trifecta delivers with free shipping to all 50 states.

Pete's Paleo

Created by Pete and Sarah Servold, this meal kit service ships nationwide and gets all their produce from southern California, where the company is based. If you've decided to cut out honey and maple syrup, you're in luck, because Pete's only uses fruit or vegetable sweeteners.

Paleo On The Go

If you're on AIP Paleo, this is the service to check out. They also have other restriction options, like Strict 30 meals that have no carrageenan, honey, maple syrup, or sulfites. They ship to all 50 states and Puerto Rico.

Green Chef

Though not exclusively Paleo, this meal kit service has a Paleo meal plan option, so when you specify that's the kit you want, all the ingredients are Paleo-friendly. You can choose a two-person or family plan. Paleo is the more expensive plan at about $15 per meal. Their packaging is all eco-friendly.

Sun Basket

With a Paleo plan meal option priced at $11.99 (for 2-4 people) or $9.99 (4 people) per serving, this meal kit service is a pretty good deal. You can skip weeks if you want without canceling your subscription. The classic menu consists of 3 recipes per week, while the family menu lets you select 2, 3, or 4 recipes.

How to make the Paleo lifestyle work for you

A diet wouldn't be very effective if it wasn't somewhat flexible. Everyone's body and needs are different, and the Paleo lifestyle is customizable to fit whoever wants to try it. There are seven "styles" of the Paleo lifestyle:

Basic Paleo

This is the bare-bones Paleo, so no grains, soy, dairy, refined, or processed foods. Honey and other "questionable" items are usually not allowed, and a person on the Basic Paleo lifestyle will probably try to include other Paleo activities in their lives, like more time outside, less technology, and a "caveman" sleep routine.

80/20 Paleo

With this diet, you are Paleo 80% of the time, and get to eat whatever you want 20% of the time. This is a good option if you aren't trying to meet any specific health goals, or if you want to support someone who is going Paleo. It's also a good choice if you're trying to ease into going totally Paleo.

Autoimmune Paleo

Abbreviated as AIP, this pseudo-Paleo lifestyle eliminates foods associated with inflammation. It's especially good for people who have fibromyalgia, lupus, IBS, Crohn's, and other autoimmune disorders. Eliminated foods include potatoes, bell peppers, tomatoes, seeds, nuts, and eggs.

The Primal Diet

The Paleo and Primal diet are very similar. The Primal diet was described in 2009 and is based on "The Primal Blueprint," a book by Mark Sisson, a former athlete and creator of the Mark's Daily Apple blog. On this diet, you are allowed to eat organic, raw dairy products, a few legumes, and fermented soy products. Both diets have the same health goals, like weight loss and a "cleaner" body.

The Pegan Diet

This is an extremely restrictive diet, because it requires you to be both Paleo and vegan. That means all your protein needs to come from plant-based sources. This is not recommended by experts on Paleo, because its benefits are due to the animal proteins and fats.

Paleo for food allergies

If you struggle with food allergies, a Paleo lifestyle can be a good choice. It naturally eliminates a lot of foods that cause allergies, like peanuts, soy, and gluten. Depending on your allergy, the

Paleo lifestyle can also be personalized to cut out shellfish, eggs, nuts, and coconut. The lack of preservatives and artificial ingredients can also provide a lot of relief for those with allergies.

The Ketogenic Paleo

This diet combines Paleo with the Keto diet, which is all about getting the body to burn fat instead of carbs for energy. That requires eating very few carbs and lots of healthy fat. While on a regular Keto diet, that would mean eating full-fat dairy, when you add Paleo, all dairy is eliminated. However, it's still possible to enter "ketosis" on a Paleo lifestyle. The Ketogenic Paleo is used by people who are very heavy and wanting to lose weight, bodybuilders, or people with diabetes. It's also used for people with epilepsy.

YOUR BODY AND TRANSITIONING INTO PALEO

There are two ways that people approach going Paleo: they slowly transition, or they go cold turkey on grains, dairy, etc, and dive right in. What works depends on who you are, but if you want to avoid feeling miserable and grumpy, a slower transition is better. However, on the other hand, going 100% Paleo right away will generate faster results.

What happens to your body during transition

Whether you go Paleo over time or all-in right away, your body will experience changes. If you go 100% Paleo on day 1, the changes will be significantly more apparent. You will probably experience what's known as "the carb flu."

If you have a typical diet, your body is used to lots of grains and not much fat or protein. That means your body relies on carbs for energy. When you go on Paleo, your body starts to switch to protein and fat. Your blood sugar also drops to a normal level while your insulin levels work to regulate themselves. When your insulin levels are regulated, your body releases triglycerides - the body's storage of fat - for energy. This is good news, because your body prefers burning fat. Fat burns more slowly than carbs, so it's more reliable and efficient.

As your body goes through these processes, it can trigger flu-like symptoms. You might feel fatigued, headaches, stomach cramps, nausea, and even "brain fog."

HOW TO COMBAT THE "CARB FLU"

The carb flu can be unpleasant and disrupt your life. Luckily, there are ways you can combat the symptoms and feel better during your body's switch from carbs to protein and fat. Here are some tips:

Stay hydrated

Drinking enough water is essential when you're transitioning to Paleo. The body is working overtime and needs fluid. Odds are, if you're like most people, you aren't drinking enough water anyway, so making that a priority can only help.

Drink bone broth mixed with salt

When you're on a typical diet, you're getting lots of sodium from processed food. When you switch to Paleo, there is not salt, so you need to remember to add it in. Salt is crucial for replenishing electrolytes. You can tell your body is low when you get headaches or feel tired. One of the best ways to treat those symptoms is to drink bone broth with salt.

Eat more

One of the easiest ways to fight the fatigue part of the carb flu is to eat more food. When you were relying on carbs, you didn't need to eat as much to feel full, but when you're eating primarily meat and vegetables, it takes much larger portions to feel the same level of satisfaction. Add more protein, fats like avocado, and tons of veggies to your meals and snacks to feel full and provide your body with enough energy.

Speed up the digestive process

When your body ingests more fiber, protein, and fat, it might need some extra help to really get going. Eating fermented foods like sauerkraut and kimchi can provide probiotics and enzymes that aid the digestive process. You can also help speed up the detoxing process by eating ginger, asparagus, and onions, which the liver and kidneys love. The liver regulates body fuel, while the kidneys are responsible for regulating electrolytes.

Add more carbs

The Paleo lifestyle is not a "low-carb" diet; it just cares about where the carbs are coming from. If you are feeling weak and miserable because you aren't getting carbs from grains, add more slow-burning carbs like squash, sweet potatoes, and cauliflower. You can also get some carbs from fruit that is lower in sugar, like oranges, lemons, limes, and berries.

Don't exercise so hard

Exercise is important to a healthy lifestyle, but while you're transitioning to burning protein and fat as fuel instead of carbs, you might want to take it easy. It's all right if you need to lower the intensity of your workouts for 2-3 weeks while your body adjusts.

TIPS FOR OVERALL SUCCESS

You know what the Paleo lifestyle is, how to stock your kitchen, and how to manage the difficult first weeks, but you can never have enough tips for success. Here are some to commit to memory and turn into daily habits. It will make transitioning and sticking to a Paleo lifestyle much easier:

Don't be afraid of fat

Since the 1980's, fat has been given a bad name, when in reality, it's sugar that's the problem. Everywhere you look, it's low-fat and non-fat. You want to avoid those products like the plague. Instead, embrace animal and fish fat, and the fat found in coconut, seeds, nuts, and avocadoes.

Don't go crazy with Paleo goodies

There are lots of goodies you can make using all Paleo ingredients, which technically makes them Paleo. However, they're still not "healthy," and you should treat them the same as you would any dessert. In other words, eat them infrequently and make them a special - not regular - treat.

Keep it simple

Being on a Paleo lifestyle requires a lot of home-cooking, which you may not be used to. Keeping recipes as simple and easy as possible is the best way to go, especially when you're just starting out. If you feel up to it, by all means experiment and try out harder recipes, but the Paleo lifestyle is really all about using a few essential ingredients to their full potential.

Always meal prep

When most of your meals are homemade, that means prepping is going to be your best friend. When you buy a bunch of veggies, prep them all right away - wash, cut, and put in containers for the week. Make a huge stew on a Sunday. Freeze sauces, stocks, and herbs. Always have Paleo snacks with you in case you get hungry. By prepping, you take out most of the hard work, so when you're exhausted and come home, you have something ready to go.

Find a support group

Making a dramatic diet change is difficult, and it's even more difficult if you're on your own. If your whole family isn't participating, find a friend who is, or find a group. There are countless options online where people share their experiences and recipes. Just knowing that there are others going through the same thing as you can make all the difference.

Get enough sleep

Sleep is absolutely essential to a healthy mind, body, and spirit. Most people don't get enough quality sleep. We sleep with our phones, with lights, and other distractions. You want your bedroom to be an oasis and a place where you're totally unplugged. Most people need at least 8 hours of sleep to reach their full potential, and if you're battling a nasty case of the carb flu, you might need more.

Think about taking supplements

When you eliminate food groups like dairy, you might be missing out on some important nutrients like Vitamin D, K2, and so on. You can get most vitamins and minerals from food, but if you're still concerned, you can get high-quality supplements.

Embrace Paleo as a lifestyle

Going Paleo is more than a diet. It's about simplifying and enjoying life in its pure form. Go on more hikes. Get rid of all the "stuff" in your house you don't need. Eliminate external sources of stress and negativity. It starts with food, but Paleo has the power to transform your entire life.

Paleo Breakfast Recipes

Contents

Scrambled Eggs with Fruit

Serves: 1 / Preparation time: 2 minutes / Cooking time: 4 minutes

This is a quick, easy and healthy breakfast, that you will enjoy making and having regularly on weekday mornings when you are press for time.

3 eggs

1 teaspoon olive oil

1/3 red tomatoes (chopped fine)

½ cup pineapple (diced)

¼ avocado (seeds and skin discarded, diced)

- Heat the oil in a pan and stir fry the egg and tomatoes in it, until done.
- Plate the scrambled egg with the avocado and pineapple.

Per Serving: Calories: 373.7; Total Fat: 25.8g; Saturated Fat: 6.3g; Protein: 20.5g; Carbs: 17.1g; Fiber: 4.5g; Sugar: 8.7g;

Grapefruit & Carrot Smoothie

Serves: 2 / Preparation time: 5 minutes / Cooking time: 2minutes

This grapefruit and carrot smoothie is a nutritious breakfast idea, when you want to make up for your fruit and veggies intake in a quick whip.

1 cup carrot (grated)

2 cups mango (frozen)

1 cup grapefruit (sliced)

1 banana (peeled)

1 cup almond milk

1 tablespoon honey

1 tablespoon ground flaxseed

- Combine all the ingredients in a high speed blender.
- Blend until smooth.

Per Serving: Calories: 277; Total Fat: 3g; Saturated Fat: 0g; Protein: 4g; Carbs: 62g; Fiber: 7g; Sugar: 40g;

Hazelnut & Chocolate Cereal

Serves: 8 / Preparation time: 10 minutes / Cooking time: 15 minutes

When you are on Paleo, breakfast cereal is something you may miss. But having this hazelnut and chocolate cereal, you will not miss it anymore.

¼ cup almond flour

1/3 cup honey

1 cup sunflower seeds (unsalted)

1 cup raw coconut (shredded)

1/3 cup cocoa powder (unsweetened)

1 cup hazelnuts (unsalted, chopped)

2 tablespoon almond butter (melted)

¼ tablespoon coconut oil

- Grease a pan with coconut oil and toss all the ingredients in it, spreading and pressing it well.
- Bake in an oven preheated to 350 degrees Fahrenheit/ 175 degrees Celsius for 15 minutes.
- Leave to cool and then crumble it.
- Serve with some coconut milk.

Per Serving: Calories: 330.8; Total Fat: 27.7g; Saturated Fat: 8gProtein: 7.6g; Carbs: 19.9g; Fiber: 6g; Sugar: 10.7g;

Milky Pancakes

Serves: 8 / Preparation time: 6 minutes / Cooking time: 14 minutes

Pancakes make quite a filling breakfast. These milky pancakes, sweetened with maple syrup taste wonderful.

1/3 cup almond flour	3 eggs
1/3 cup coconut flour	1/8 teaspoon sea salt
1 teaspoon vanilla extract	2 teaspoon pure maple syrup
1 tablespoon coconut oil (melted)	½ cup almond milk
¼ teaspoon baking soda	½ teaspoon apple cider vinegar

- Combine the almond milk and apple cider vinegar in a bowl and place aside for 2 minutes.
- Mix together all the ingredients in a bowl including the prepared buttermilk.
- Heat a lightly greased skillet on medium flame and spoon some batter onto it.
- Cook until bubbles are formed on top and then turn the pancake, cooking until the center is done.
- Repeat with the rest of the batter.

Per Serving: Calories: 85; Total Fat: 6g; Saturated Fat: 1.3g; Protein: 3g; Carbs: 5g; Fiber: 2g; Sugar: 2g;

Kale & Sausage Muffins

Serves: 12 / Preparation time: 5 minutes / Cooking time: 25 minutes

Muffins are a delightful breakfast and when made with breakfast sausage and some veggies it's something you definitely can't resist.

9 eggs (whisked)

8 oz ground chicken breakfast sausage

¼ teaspoon pepper

½ cup kale (chopped)

1 red pepper (chopped)

2 tablespoon olive oil

- Brown the sausage in a skillet and place in a bowl along with the remaining ingredients and mix well.
- Pour the batter into greased muffin-tin and bake in an oven preheated to 350 degrees Fahrenheit/ 175 degrees Celsius for 20-25 minutes.
- Leave to cool.

Per Serving: Calories: 105; Total Fat: 7.6g; Saturated Fat: 2g; Protein: 7.7g; Carbs: 1.3g; Fiber: 0.2g; Sugar: 0.8g;

Banana Pancakes

Serves: 3 / Preparation time: 5 minutes / Cooking time: 6 minutes

Serving these banana pancakes with a sprinkle of honey and some almond butter make a sumptuous breakfast.

5/8 cup almond flour

¼ teaspoon baking soda

2 eggs

¼ teaspoon vanilla extract

½ cup coconut milk (whole fat)

1/8 cup coconut flour

¼ cup banana (mashed)

½ teaspoon ground cinnamon

¼ teaspoon sea salt

- Mix together all the dry ingredients and wet ingredients in separate bowls.
- Heat a lightly greased skillet on medium flame and spoon some batter onto it.
- Cook until bubbles are formed on top and then turn the pancake, cooking until the center is done.
- Repeat with the rest of the batter.

Per Serving: Calories: 300.1; Total Fat: 22.5g; Saturated Fat: 7.5g; Protein: 5.9g; Carbs: 15.1g; Fiber: 5g; Sugar: 5.9g;

Pumpkin Spiced Puree

Serves: 1 / Preparation time: 10 minutes / Cooking time: 7 minutes

The flavor of pumpkin spice is a trendy favorite when the weather gets cooler, but there is no reason why you can't enjoy this meal any time of the year. This Paleo breakfast resembles a sort of a pudding. Top it with some berries to add to the taste.

4 egg whites

½ cup pumpkin puree

1 teaspoon vanilla extract

1 teaspoon cinnamon

1 teaspoon pumpkin pie spice

Pinch of sea salt

2 tablespoon flaxseed meal (ground)

- Whisk the egg whites and then mix in the remaining ingredients.
- Place the mixture in a greased saucepan over medium flame and stir cook for 5-6 minutes.
- Served garnished with berries.

Per Serving: Calories: 199.5; Total Fat: 5.3g; Saturated Fat: 0.1g; Protein: 21.3g; Carbs: 18.5g; Fiber: 10.9g;

Spiced Pumpkin Pancakes

Serves: 2 / Preparation time: 5 minutes / Cooking time: 5 minutes

These crispy mushroom pops with a moist interior are just mouthwatering. It makes a wonderful party time appetizer that guests will relish.

1/2 cup canned pumpkin

1 tablespoon honey

½ teaspoon cinnamon

¼ teaspoon nutmeg

¼ teaspoon vanilla

2 eggs

½ cup almond butter

- Mix together all the ingredients in a bowl.

- Heat a lightly greased skillet on medium flame and spoon some batter onto it.

- Cook until edges become stiff and then turn the pancake, cooking until the center is done.

- Repeat with the rest of the batter.

Per Serving: Calories: 528.9; Total Fat: 42.5g; Saturated Fat: 5.3g; Protein: 16.4g; Carbs: 28.7g; Fiber: 5.2g; Sugar: 11g;

Chia Bran Bread

Serves: 2 / Preparation time: 2 minutes / Cooking time: 2 minutes

When you have absolutely no time to spend in the kitchen on a busy morning, this microwave bread can be of great help. Some argue that bread can never be Paleo, while others feel that if all of the ingredients are Paleo-friendly , then bread is fine to eat. It is up to you to decide whether you wish to include Paleo-friendly bread in your diet.

1 tablespoon chia bran

1/8 teaspoon sea salt

½ teaspoon Paleo baking powder

1/3 cup almond flour

2 ½ tablespoon coconut oil

1 egg (whisked)

- Whisk together all the ingredients and then pour it into a greased mug.
- Microwave for 1 ½ minute on high.
- Leave to cool and then remove and slice.

Per Serving: Calories: 305.5; Total Fat: 28.9g; Saturated Fat: 7.9g; Protein: 4g; Carbs: 5.7g; Fiber: 3g; Sugar: 0.8g;

Nutty Banana Smoothie

Serves: 1 / Preparation time: 4 minutes / Cooking time: 0 minutes

You can substitute the bananas with other fruits as well like berries. You do not necessarily need to use dairy.

1 large ripe banana   ¼ cup almonds (chopped)

4 ice cubes

- Combine all the ingredients in a blender.
- Blend until smooth.

Per Serving: Calories: 258; Total Fat: 12.2g; Saturated Fat: 1.1g; Protein: 6.5g; Carbs: 36.2g; Fiber: 6.5g; Sugar: 17.6g;

Flax Meal Cereal

Serves: 1 / Preparation time: 5 minutes / Cooking time: 3 minutes

On a cold morning, there are times all you want is a steaming bowl of cereal. This flax meal cereal will save your day.

2 tablespoon flax meal

2 tablespoon chia seeds

2 tablespoon unsweetened coconut (shredded)

½ cup boiling water

¼ cup almond milk (unsweetened)

½ tablespoon cinnamon

- Combine all the dry ingredients in a bowl.

- Pour in the water and then stir mix, leaving aside for 3 minutes.

- Stir yet again and mix in the almond milk.

Per Serving: Calories: 282.2; Total Fat: 18.2g; Saturated Fat: 6.5g; Protein: 10g; Carbs: 19.9g; Fiber: 17.4g; Sugar: 0.7g;

Beef Muffins

Serves: 6 / Preparation time: 10 minutes / Cooking time: 35 minutes

A breakfast dish that is extremely simple to prepare yet amazingly tasty. You can use different vegetables or meat to try out some variations in the dish.

1 cup extra-lean beef sirloin (ground)   1 red bell pepper (chopped fine)

2 scallions (chopped)

6 eggs (whisked)

- Brown the beef in a skillet.

- Mix together all the ingredients well and pour into a 6-well muffin tin.

- Bake in an oven preheated to 350 degrees Fahrenheit/ 175 degrees Celsius for 35 minutes.

Per Serving: Calories: 132.1; Total Fat: 8.1g; Saturated Fat: 2.7g; Protein: 12.9g; Carbs: 1.7g; Fiber: 0.5g; Sugar: 0.1g;

Green Mushroom Quiche

Serves: 6 / Preparation time: 10 minutes / Cooking time: 30 minutes

This green mushroom breakfast quiche is a great breakfast dish to prepare on the weekends especially when you have guests at home.

15 eggs

½ onion (chopped)

3 garlic cloves (minced)

1 ½ teaspoon Paleo baking powder

3 cups fresh spinach (chopped)

5 mushrooms (chopped)

Sea salt to taste

Ground black pepper to taste

1 ½ cup coconut milk

- Whisk together all the ingredients and then pour the mixture into a greased baking dish.
- Bake in an oven preheated to 350 degrees Fahrenheit/ 175 degrees Celsius for around 30-40 minutes.

Per Serving: Calories: 193.6; Total Fat: 13.6g; Saturated Fat: 6.5g; Protein: 13.4g; Carbs: 3.7g; Fiber: 0.7g; Sugar: 1.3g;

Choco-Zucchini Bread

Serves: 8 / Preparation time: 10 minutes / Cooking time: 35 minutes

This choco-banana bread is power packed with nutrients. It will leave you energetic as you begin your day. It is up to you whether you wish to include Paleo-friendly breads in your diet, or whether you decide to go without bread entirely.

½ cup almonds (ground)

1 cup zucchini (grated)

1 banana

2 eggs

1 teaspoon Paleo baking powder

1 teaspoon raw honey

1 teaspoon cinnamon (ground)

2 pieces Ghirardelli chocolate (86% cocoa)

- Whisk together all the ingredients and then pour the mixture into a greased bread dish.
- Bake in an oven preheated to 350 degrees Fahrenheit/ 175 degrees Celsius for around 30-35 minutes.

Per Serving: Calories: 85; Total Fat: 5.8g; Saturated Fat: 1.5g; Protein: 3.1g; Carbs: 7.1g; Fiber: 1.8g; Sugar: 3.3g;

Breakfast Fritters

Serves: 2 / Preparation time: 10 minutes / Cooking time: 8 minutes

Breakfast fritters can be a nice change from the same old breakfast routine. Not only that, this recipe is very healthy so you will get your day started on the right foot.

2 cups zucchini (shredded)

3 eggs

½ teaspoon sea salt

¼ teaspoon ground black pepper

1 tablespoon coconut flour (sifted)

- Whisk together all the ingredients in a bowl.
- Heat a lightly greased skillet on medium flame and spoon some batter onto it.
- Cook until set and then turn the fritter, cooking until the center is done.
- Repeat with the rest of the batter.

Per Serving: Calories: 151.2; Total Fat: 7.4g; Saturated Fat: 2g; Protein: 10.3g; Carbs: 12.1g; Fiber: 5.5g; Sugar: 3g;

Paleo Soups, Salads & Dressing Recipes

Contents

Egg & Tuna Salad

Serves: 1/ Preparation time: 10 minutes / Cooking time: 0 minutes

This tuna and egg salad is quite a filling meal in itself. You can have this for a power packed lunch.

2 cups red leaf lettuce (torn)

1/2 carrot (chopped)

1 tablespoon olive oil

1 celery stalk (chopped)

1 1/4 oz mango (chopped)

1 Boiled Egg (peeled, chopped)

1/2 cup red tomatoes (chopped)

5 oz canned tuna in water
(w/o salt)

- Toss together all the ingredients in a bowl.

Per Serving: Calories: 739; Total Fat: 22.2g; Saturated Fat: 4.4g; Protein: 115.4g; Carbs: 14.7g; Fiber: 2.7g; Sugar: 9.3g;

Watercress & Cauliflower Soup

Serves: 10 / Preparation time: 10 minutes / Cooking time: 30 minutes

On a cold day, when all you want is to get cozy in your sofa, this watercress and cauliflower soup will surely keep you warm.

1 tablespoon almond butter

1 white onion (chopped)

5 celery stalks (chopped)

2 Tbsp lemon Juice

2 Tsp coriander

2 Tsp basil

32 oz organic chicken stock

1 ½ Cups of Radishes (chopped)

1 Cup Cauliflower (chopped)

13 1/2 oz organic coconut milk

6 Cups watercress leaves

3/4 Tsp Sea Salt

1 tablespoon garlic powder

- Heat almond butter in a pot and sauté the herbs in it.
- Add the rest of the ingredients except the watercress and cook until the cauliflower and radish is tender.
- Mix in the watercress and leave to simmer until wilted.
- Leave to cool a while and then puree the soup in a blender.
- Transfer back into the pot and simmer for another 5-10 minutes.

Per Serving: Calories: 120; Total Fat: 10.4g; Saturated Fat: 8.3g; Protein: 2.8g; Carbs: 5.8g; Fiber: 2.3g; Sugar: 3.1g;

Lemon & Herb Shrimp Soup

Serves: 8 / Preparation time: 15 minutes / Cooking time: 30 minutes

If you are a fan of shrimps you will surely love this soup. You can serve this soup topped with some red pepper flakes for extra spice.

2 cans of tomatoes

17 ½ oz shrimps (peeled, deveined)

2 ½ cups coconut milk

¼ cup cilantro (chopped)

¼ cup chives (chopped)

6 garlic cloves (minced)

2 tablespoon coconut oil

2 hot chili peppers

3 lemons (juiced)

4 onions (chopped)

- Heat oil in a pot and cook the onion, tomatoes, garlic, chives, cilantro and pepper.
- Cook covered for 20 minutes on low flame.
- Cool and then blend in a blender in batches.
- Transfer back to the pot, mix in the remaining ingredients and cook for another 10 minutes.

Per Serving: Calories: 301.6; Total Fat: 16.3g; Saturated Fat: 10.8g; Protein: 17.2g; Carbs: 24.5g; Fiber: 4.8g; Sugar: 2.3g;

Taco Chicken Soup

Serves: 6 / Preparation time: 10 minutes / Cooking time: 6 hours

This taco chicken soup is low in carbohydrates and is great for those who are looking to cut off those extra carbs.

3 chicken breasts (skinless, boneless)

10 teaspoon taco seasoning

1 can tomatoes (diced)

1 spaghetti squash (halved)

- Bake the squash in an oven preheated to 350 degrees Fahrenheit/ 175 degrees Celsius.
- Remove the flesh of the squash using a fork and place in a slow cooker along with the rest of the ingredients.
- Cook for 6-8 hours on low.

Per Serving: Calories: 194.7; Total Fat: 3.4g; Saturated Fat: 0.5g; Protein: 28g; Carbs: 12.1g; Fiber: 2.1g; Sugar: 4g;

Herb & Garlic Dressing

Serves: 4 / Preparation time: 5 minutes / Cooking time: 0 minutes

You can have these lemon flavored roasted green beans as a side dish to a variety of meals. These beans are not only nutritious, but the lemon adds a nice flavor as well.

2 tablespoon lemon juice

1 teaspoon parsley (chopped fine)

1 teaspoon thyme leaves (chopped fine)

Sea salt and pepper to taste

2 tablespoon lemon juice

1 garlic clove (minced)

- Mix together all the ingredients in a bowl.

Per Serving: Calories: 64; Total Fat: 7.1g; Saturated Fat: 1.1g; Protein: 0.1g; Carbs: 0.6g; Fiber: 0.2g; Sugar: 0.2g;

Ham Veggie Soup

Serves: 8 / Preparation time: 15 minutes / Cooking time: 2 hours

The bone-in ham lends a wonderful taste to the soup. The veggies in the soup add to your nutrient intake.

1 fresh ham (bone-in, meat chopped)

½ onion (chopped)

8 cups water

2 cups ham (chopped)

2 cans green beans

1 tablespoon garlic (chopped)

16 oz cauliflower (chopped)

1 tablespoon almond butter

Sea salt to taste

Black pepper to taste

- Place all the ingredients in a pot and bring to boil.
- Simmer covered for 1 – 1 ½ hour.
- Discard the bone, retaining any marrow or meat aside.
- Using an immersion blender, puree the soup.
- Mix in the reserved meat.

Per Serving: Calories: 68; Total Fat: 4.2g; Saturated Fat: 1.1g; Protein: 4.1g; Carbs: 4.4g; Fiber: 1.8g; Sugar: 1.8g;

Spicy Shrimp Soup

Serves: 3 / Preparation time: 20 minutes / Cooking time: 15 minutes

The spice in this soup can be adjusted according to your preference by reducing the quantity of jalapeno peppers and adding an extra bell pepper. Some people also enjoy adding a raw sliced cucumber after cooking to reduce the heat.

1 pound shrimps (peeled, deveined)

1 can coconut milk

1 onion (chopped)

1 can tomatoes (diced)

1 yellow bell pepper (chopped)

2 jalapeno peppers (chopped)

2 garlic cloves

¼ cup cilantro (chopped)

2 tablespoon sriracha

- Mix together the jalapenos, bell pepper and onion in a pot and sauté.
- Add in the tomatoes, garlic, shrimps and cilantro, stir cooking until the shrimps are opaque.
- Mix in the sriracha and coconut milk and cook until done.

Per Serving: Calories: 479.1; Total Fat: 27.3g; Saturated Fat: 22.3g; Protein: 36g; Carbs: 22.9g; Fiber: 4.7g; Sugar: 7.3g;

Zucchini Salad

Serves: 6 / Preparation time: 10 minutes / Cooking time: 0 minutes

This quick summer salad is quite refreshing when served cold. It is the perfect, healthy, Paleo salad.

5 cups zucchini (julienned)

¼ cup basil (chopped fine)

1 garlic clove (chopped)

1 cup cherry tomatoes (quartered)

¼ cup extra virgin olive oil

Black pepper to taste

Sea salt to taste

- Toss together all the ingredients in a salad bowl.

Per Serving: Calories: 110; Total Fat: 9.2g; Saturated Fat: 1.2g; Protein: 1.2g; Carbs:7.3g; Fiber: 2.5g; Sugar: 2.5g;

Balsamic Mustard Dressing

Serves: 10/ Preparation time: 5 minutes / Cooking time: 0 minutes

This Italian dressing not only tastes great over salads but can also be used as a marinade for meats.

1 garlic clove (crushed)

1 teaspoon dried oregano

¼ cup balsamic vinegar

¾ cup extra-virgin olive oil

2 teaspoon Dijon mustard

Ground black pepper to taste

Sea salt to taste

- Combine all the ingredients in a jar.
- Close it and shake well.

Per Serving: Calories: 132; Total Fat: 15.2g; Saturated Fat: 2.2g; Protein: 0.1g; Carbs: 0.3g; Fiber: 0.1g; Sugar: 0g;

Lemon & Mustard Dressing

Serves: 10 / Preparation time: 5 minutes / Cooking time: 0 minutes

This classic dressing goes well with just about any dish. It simply tastes wonderful over smoked fish and no doubt a perfect dressing for salads.

¾ cup extra-virgin olive oil

½ teaspoon Dijon mustard

3 tablespoon lemon juice

Ground black pepper to taste

Sea salt to taste

- Combine all the ingredients in a jar.
- Close it and shake well.

Per Serving: Calories: 131; Total Fat: 15.2g; Saturated Fat: 2.2g; Protein: 0.1g; Carbs: 0.1g; Fiber: 0g; Sugar: 0.1g;

Asian-Style Dressing

Serves: 10 / Preparation time: 5 minutes / Cooking time: 0 minutes

This Asian Style dressing is wonderful when drizzled over roasted veggies, and salads. The ginger lends a nice taste to the dish.

2/3 cup extra-virgin olive oil

1 tablespoon fresh ginger juice

Ground black pepper to taste

Sea salt to taste

3 tablespoon rice vinegar

- Combine all the ingredients in a jar.
- Close it and shake well.

Per Serving: Calories: 120; Total Fat: 13.5g; Saturated Fat: 1.9g; Protein: 0.1g; Carbs: 0.4g; Fiber: 0.1g; Sugar: 0g;

Citus & Honey Dressing

Serves: 10 / Preparation time: 5 minutes / Cooking time: 0 minutes

This dressing is an amazing amalgamation of citrus, honey and herbs. You can have this dressing over salads and fruits.

1/4 cup extra-virgin olive oil

1 tablespoon orange zest

3 tablespoon fresh orange juice

Ground black pepper to taste

Sea salt to taste

1 tablespoon parsley (minced)

2 teaspoon raw honey

- Combine all the ingredients in a jar.
- Close it and shake well.

Per Serving: Calories: 50; Total Fat: 5.1g; Saturated Fat: 0.7g; Protein: 0.1g; Carbs: 1.8g; Fiber: 0.1g; Sugar: 1.5g;

Watermelon & Grapefruit Salad

Serves: 4 / Preparation time: 15 minutes / Cooking time: 0 minutes

An extremely refreshing salad to beat the summer heat. The fruit, the greens and the nuts make for a tasty Paleo meal.

2 cups watermelon (chopped)

1 grapefruit (peeled, segmented)

¼ cup walnuts (chopped)

4 cups mixed salad greens

½ cup strawberries (sliced)

- Toss together all the ingredients in a bowl.

Per Serving: Calories: 109; Total Fat: 4.9g; Saturated Fat: 0.3g; Protein: 4.4g; Carbs: 14.9g; Fiber: 1.5g; Sugar: 7.9g;

Spinach & Pear Salad

Serves: 2 / Preparation time: 15 minutes / Cooking time: 12 minutes

Salads do not necessarily have to be chilled. On a winter day, a warm salad will be an absolute delight.

10 mushrooms (sliced)	¼ cup almonds (chopped)
2 pears (diced)	Sea salt to taste
5 cups baby spinach	Ground black pepper to taste
½ red onion (sliced)	2 tablespoon olive oil
¼ cup balsamic vinegar	1 tablespoon fresh parsley (chopped)

- Heat oil in a skillet and sauté the onion, mushrooms and pears in it for 4-5 minutes.
- Toss in the salt, pepper and vinegar.
- Mix in the spinach and cook for 1-2 minutes.
- Serve topped with parsley and nuts.

Per Serving: Calories: 364; Total Fat: 20.8g; Saturated Fat: 2.5g; Protein: 8.6g; Carbs: 43.1g; Fiber: 11.2g; Sugar: 24.1g;

Simple Onion & Tomato Salad

Serves: 2 / Preparation time: 15 minutes / Cooking time: 0 minutes

This simple onion and tomato salad is a wonderful side dish to serve with grilled fish and meat. You can substitute the onion for green onions if you prefer.

1 onion (sliced)

3 tablespoon virgin olive oil

2 cups grape tomatoes (quartered)

2 tablespoon fresh basil (minced)

Sea salt to taste

Ground black pepper to taste

2 tablespoon balsamic vinegar

- Toss together all the ingredients in a salad bowl.

Per Serving: Calories: 238; Total Fat: 21.4g; Saturated Fat: 3.1g; Protein: 2.3g; Carbs: 12.4g; Fiber: 3.4g; Sugar: 7.1g;

Paleo Poultry Recipes

Contents

Spicy Chicken Chili

Serves: 6 / Preparation time: 10 minutes / Cooking time: 4 hours

This chicken chili spiced with jalapenos and herbs exude an amazing flavor when cooked using a slow cooker.

2 ½ pounds chicken breast (boneless, skinless)

1 onion (minced)

4 garlic cloves (minced)

2 jalapenos (diced)

2 poblano peppers (diced)

½ cup cilantro (chopped)

6 cups chicken broth (low-sodium)

Sea salt to taste

Ground black pepper to taste

2 teaspoon cumin

1 teaspoon oregano

1 lime

- Combine all ingredients in the slow cooker and cook for 4 hours on low.
- Shred the chicken and mix it back into the cooker mixture.

Per Serving: Calories: 279; Total Fat: 6.4g; Saturated Fat: 0.4g; Protein: 45.9g; Carbs: 6.7g; Fiber: 1.3g; Sugar: 2.7g;

Hot Chicken

Serves: 6 / Preparation time: 10 minutes / Cooking time: 4 hours

This one-pot chicken dish is absolutely mouth-watering. Serve it to your guests and they will truly enjoy it.

2 ½ pounds chicken breast (boneless, skinless)

1 onion (minced)

20 oz fire roasted crushed tomatoes

3 tablespoon fire-roasted green chilies (diced)

1 teaspoon oregano

1 teaspoon garlic powder

½ teaspoon coriander

½ teaspoon cumin

Salt and pepper to taste

- Season the chicken with salt and pepper.
- Add the chicken to the crock pot with the rest of the ingredients.
- Cook for 4 hours on low.
- Shred the chicken and mix it back into the cooker mixture.

Per Serving: Calories: 263; Total Fat: 4.8g; Saturated Fat: 0g; Protein: 40.4g; Carbs: 9.7g; Fiber: 2g; Sugar: 5.1g;

Chicken & Mushroom Toss

Serves: 4/ Preparation time: 15 minutes / Cooking time: 1 hour 45 minutes

This is a savory dish to prepare at family get-togethers. Not only is it fast and easy to prepare, it is always a hit with groups.

1 ½ pounds chicken breasts (boneless, skinless)

8 oz mushrooms (sliced)

2 garlic cloves (minced)

Ground black pepper to taste

½ cup chicken broth (low-sodium)

2 ½ tablespoon balsamic vinegar

½ teaspoon thyme

1 tablespoon parsley (chopped)

2 teaspoon olive oil

Sea salt to taste

- Season the chicken with salt and pepper.
- Heat oil in a skillet and sear the chicken pieces in it on both sides for 2-3 minutes. Place aside.
- Add the mushrooms and garlic to the skillet and cook for around 3-4 minutes.
- Add the rest of the ingredients including the chicken, deglaze the bottom and leave to simmer for 10-15 minutes.

Per Serving: Calories: 292; Total Fat: 12.3g; Saturated Fat: 3.1g; Protein: 41.4g; Carbs: 2.2g; Fiber: 0.6g; Sugar: 0.9g;

Turkey Stir Fry

Serves: 4 / Preparation time: 10 minutes / Cooking time: 20 minutes

This turkey stir fry is something that everyone will love. It isn't spicy so it can be a great dish for kids or those who just don't like much heat.

2 tablespoon olive oil

2 garlic cloves (minced)

2 tablespoon ginger (minced)

1 pound ground turkey (99% lean)

4 oz Brussels sprouts

4 oz carrots (chopped)

4 oz kale (torn)

4 oz cabbage (shredded)

10 teaspoon coconut aminos

2 tablespoon rice vinegar

- Heat the oil in a skillet and sauté the garlic, ginger and turkey in it until fully cooked.

- Add the veggies and cook for 4-5 minutes.

- Mix in the coconut aminos and rice vinegar and cook for a minute.

Per Serving: Calories: 355; Total Fat: 19.8g; Saturated Fat: 3.2g; Protein: 33.8g; Carbs: 14.9g; Fiber: 3.3g; Sugar: 3g;

Chicken Stew

Serves: 4 / Preparation time: 20 minutes / Cooking time: 8 hours

If you want to cook a large meal, then this chicken stew is something you need to try out.

4 chicken thighs (skinless, boneless)

1 onion (chopped)

4 cups baby carrots

1 cauliflower head (chopped)

3 zucchini (chopped)

1 tablespoon fresh ginger

2 cups of canned diced tomatoes

1 lime (juiced)

3 garlic cloves

1 tablespoon cumin seeds

1 tablespoon paprika

2 teaspoon turmeric

1 tablespoon ground cinnamon

- Grease the slow cooker with olive oil spray and mix all the ingredients in it.
- Cook for 8-10 hours on low..

Per Serving: Calories: 271.8; Total Fat: 4.3g; Saturated Fat: 0.9g; Protein: 21.5g; Carbs: 42g; Fiber: 14.1g; Sugar: 11.4g;

Baked Nutty Chicken

Serves: 4 / Preparation time: 15 minutes / Cooking time: 25 minutes

The almond flavor in this chicken bake is distinct and adds some healthy fats.

1 pound chicken breast (skinless, boneless)

1 tablespoon dried basil

1 ½ teaspoon garlic powder

½ teaspoon black pepper

½ teaspoon sea salt

¼ teaspoon paprika

1 ½ teaspoon Italian seasoning

½ cup almond meal

1 egg

1 egg white

- Line a baking pan with foil, greasing it with olive oil.

- Whisk together the egg and egg white in a bowl, and mix the dry ingredients in a separate bowl.

- First dip the chicken in egg and then coat it with the dry mixture, placing each piece in the baking dish.

- Bake for 12 minutes in an oven preheated to 400 degrees Fahrenheit/ 205 degrees Celsius.

- Flip and cook for another 12 minutes.

Per Serving: Calories: 196.8; Total Fat: 8.4g; Saturated Fat: 0.8g; Protein: 27.7g; Carbs: 4.4g; Fiber: 1.6g; Sugar: 1.1g;

Turkey & Veggie Toss

Serves: 1 / Preparation time: 10 minutes / Cooking time: 10 minutes

This turkey and veggie toss is a quick meal to prepare but is a complete meal by itself. You can also replace the turkey with chicken.

3 oz turkey breast

¼ cup onions (sliced)

½ cup mushrooms (chopped)

1 tablespoon extra-virgin olive oil

3 tablespoon water

½ cup zucchini (sliced)

3 small asparagus spears

- Heat oil in a skillet and sauté the onion in it.
- Mix in the meat and cook for 2-3 minutes.
- Add in the veggies with water and cook covered until ready.
- Add a little water if required to avoid sticking.

Per Serving: Calories: 249.8; Total Fat: 15.7g; Saturated Fat: 2.5g; Protein: 17.3g; Carbs: 12.4g; Fiber: 3.4g; Sugar: 5.2g;

Turkey Meatloaf

This turkey meatloaf is a must try. Too often turkey is regarded as a Thanksgiving meal only, then forgotten about for the rest of the year. Enjoy this healthy Paleo turkey dish any time of the year.

2 pounds ground turkey

1 egg (whisked)

2 tablespoon coconut aminos

½ cup celery (chopped)

½ cup onion (chopped)

½ cup green pepper (chopped)

Sea salt to taste

Ground black pepper to taste

½ teaspoon oregano

- Combine all the ingredients in a bowl and transfer into a loaf pan greased with coconut oil.
- Bake for 60 minutes in an oven preheated to 375 degrees Fahrenheit/ 190 degrees Celsius.

Per Serving: Calories: 127.8; Total Fat: 3.3g; Saturated Fat: 1g; Protein: 21.6g; Carbs: 2.1g; Fiber: 0.3g; Sugar: 0.9g;

Salsa Chicken Wraps

Serves: 2 / Preparation time: 5 minutes / Cooking time: 15 minutes

A dish similar to chicken fajitas, but one that is really healthy and quick to prepare with lettuce wraps replacing tortillas.

8 oz chicken breast

1 red onion (chopped)

2 red bell peppers (chopped)

¼ cup salsa

1 teaspoon olive oil

10 lettuce leaves

- Heat oil in a skillet and cook the chicken in it until cooked through.
- Place aside and cook the onion and peppers on it.
- Shred the chicken and mix it back in the skillet along with the salsa.
- Wrap the stuffing in lettuce leaves and serve.

Per Serving: Calories: 339.5; Total Fat: 15.6g; Saturated Fat: 2.3g; Protein: 30g; Carbs: 21g; Fiber: 5.9g; Sugar: 0.4g;

Citrus Chicken

Serves: 8 / Preparation time: 10 minutes / Cooking time: 8 hours

When slow cooked, the orange flavor sinks deep into the chicken adding to the aroma and also to the taste.

4 pounds chicken legs (skin removed)

1/2 cup fresh orange juice

4 garlic cloves

2 tablespoon fresh thyme

1 tablespoon fresh ginger (minced)

1 teaspoon ground allspice

4 green onions

1 ½ teaspoon sea salt

2 tablespoon white vinegar

3 red peppers (seeded, chopped)

2 cups pineapple (chopped)

- Combine the garlic, orange juice, allspice, ginger, vinegar thyme, onions, green onions, and red peppers in a food processor, blending until blended well.

- Season the chicken with salt and place in the slow cooker along with the blended mixture.

- Leave to cook for 8 hours on low.

- Add the pineapple, 30 minutes prior to the completion of cooking.

Per Serving: Calories: 475; Total Fat: 17.1g; Saturated Fat: 4.7g; Protein: 66.6g; Carbs: 10.7g; Fiber: 1.5g; Sugar: 6.5g;

Paleo Pork Recipes

Contents

Pork & Cabbage Stir Fry

Serves: 1 / Preparation time: 10 minutes / Cooking time: 15 minutes

This cabbage and pork stir fry is a quick, and tasty dish. You can always substitute the pork for other meats if you prefer.

3 oz pork shoulder (chopped)

½ cup onion (chopped)

½ cup mushrooms (chopped)

1 tablespoon olive oil

2 fl oz water

1 cup cabbage (chopped)

- Heat oil in a skillet and sauté the onions in it.
- Add the meat and stir while cooking, adding the water.
- Mix in the cabbage and mushrooms and cook for around 8-10 minutes.

Per Serving: Calories: 365.8; Total Fat: 25.1g; Saturated Fat: 8.1g; Protein: 45.9g; Carbs: 12.9g; Fiber: 3.9g; Sugar: 0.7g;

Pork & Asparagus Stir Fry

This asparagus and pork stir fry is absolutely lip-smacking. It gives you a good portion of both meat and veggies.

3 oz pork shoulder (chopped)

1 cup onion (chopped)

1 cup mushrooms (chopped)

1 tablespoon olive oil

¼ cup water

1 cup asparagus (chopped)

1 sprig rosemary

1 small lemon (sliced, for garnish)

- Heat oil in a skillet and sauté the onions in it.

- Add the meat and stir while cooking, adding the water.

- Mix in the asparagus, chopped rosemary and mushrooms and cook for around 8-10 minutes.

- Garnish with slice of lemon and serve.

Per Serving: Calories: 314; Total Fat: 12g; Saturated Fat: 2.5g; Protein: 32.1g; Carbs: 22.2g; Fiber: 6.5g; Sugar: 1.3g;

Thai Pork Curry

Serves: 2 / Preparation time: 10 minutes / Cooking time: 20 minutes

This amazing Paleo pork curry can be served over rice. You can serve these garnished with some Kaffir lime leaves.

8 oz pork loin (thinly sliced)

1 red bell pepper (chopped)

1 can coconut milk

1 cup broccoli florets

2 teaspoon green curry paste

1/3 tablespoon fish sauce

- Heat the coconut milk in a pot for 3 minutes and then mix in the curry paste.

- Add the meat and meat and fish sauce and stir cook for another 6 minutes.

- Mix in the veggies and cook for 5-10 minutes.

Per Serving: Calories: 345; Total Fat: 17.4g; Saturated Fat: 9.5g; Protein: 35.2g; Carbs: 10.6g; Fiber: 3.5g; Sugar: 3.3g;

Pork Meatballs

These pork meatballs are super delicious. When you try one, you will definitely crave some more.

1 pound ground pork

1 tablespoon apple cider vinegar

2 tablespoon of spice mix

Spice Mix:

1 tablespoon garlic powder

1 tablespoon onion powder

½ tablespoon sea salt

1 tablespoon smoked paprika

2 tablespoon chipotle powder

1 teaspoon black pepper

- Mix together all the ingredients in a bowl and shape the mixture into meatballs.

- Arrange the meatballs in a baking pan lined with aluminum foil.

- Bake in an oven preheated to 425 degrees Fahrenheit/ 218 degrees Celsius for 20-25 minutes.

Per Serving: Calories: 117.4; Total Fat: 10g; Saturated Fat: 4.1g; Protein: 6.3g; Carbs: 0.1g; Fiber: 0g; Sugar: 0.1g;

Herbed Pork Chops

Serves: 4 / Preparation time: 10 minutes / Cooking time: 40 minutes

Enjoy pork chops in a new way with this blend of herbs and spices. Throw in a chopped jalapeno pepper if you want to add some extra heat.

4 pork loin chops

½ teaspoon sea salt

¼ teaspoon paprika

¼ teaspoon ground black pepper

¼ teaspoon dried thyme

¼ teaspoon sage

1 tablespoon coconut oil

1 onion (sliced)

- Mix together all the ingredients in a bowl except the pork, onion and coconut oil and marinate the pork with the mixture.

- Heat the coconut oil in a skillet and brown both sides of the chop in it.

- Place the pork in a baking pan with foil and place the onion over it, closing the foil to totally cover the chops.

- Bake in an oven preheated to 425 degrees Fahrenheit/ 218 degrees Celsius for 30 minutes.

Per Serving: Calories: 297; Total Fat: 23.3g; Saturated Fat: 10.4g; Protein: 18.3g; Carbs: 2.8g; Fiber: 0.7g; Sugar: 1.2g;

Balsamic Pork Roast

Serves: 8 / Preparation time: 15 minutes / Cooking time: 6-8 hours

This balsamic pork roast is a party hit. Don't be put off by the long cooking time. You can prepare this dish in advance without spending too much time in the kitchen.

2 pound pork shoulder roast (boneless)

½ teaspoon garlic powder

1/3 cup chicken broth

½ teaspoon red pepper flakes

1/3 cup balsamic vinegar

1 tablespoon honey

1 tablespoon Worcestershire sauce

Sea salt to taste

- Season the pork with garlic powder, red pepper flakes and salt.

- Place the pork in the slow cooker.

- Mix together the rest of the ingredients and pour over the pork.

- Cook 6-8 hours on low.

- Remove the pork and break it apart lightly, adding it back to the slow cooker.

Per Serving: Calories: 306; Total Fat: 23.2g; Saturated Fat: 10.4g; Protein: 8.1g; Carbs: 2.9g; Fiber: 0.1g; Sugar: 2.7g;

Crispy Almond Crusted Pork

Serves: 4 / Preparation time: 10 minutes / Cooking time: 12 minutes

The rich almond flavor coating the pork chops makes this dish one you will want to make over and over. It is crispy on the outside and moist on the inside.

1 ½ pounds center cut pork loin chops (boneless)

3 tablespoon Dijon mustard

2 egg whites (whisked)

1/3 cup almond meal

½ teaspoon sea salt

½ teaspoon pepper

2 teaspoon olive oil

¾ teaspoon dried thyme

- Mix the salt, pepper, thyme and mustard.
- Coat the pork with the mustard mixture.
- Dip the pork chops first into the egg and then dredge it in the almond meal.
- Heat olive oil in a skillet and lightly fry the pork chops in it until crispy brown for 2-3 minutes per side.
- Transfer the skillet into an oven preheated to 450 degrees Fahrenheit for 9-10 minutes.

Per Serving: Calories: 294; Total Fat: 12.8g; Saturated Fat: 2.9g; Protein: 41.7g; Carbs: 2.7g; Fiber: 1.5g; Sugar: 0.6g;

Lime & Garlic Pork Chops

Serves: 4 / Preparation time: 10 minutes / Cooking time: 10 minutes

These pork chops can be made with simple ingredients that are found in your pantry. More importantly, they are ready in no time.

4 l pork chops (boneless)

4 garlic cloves (crushed)

½ teaspoon chili powder

½ teaspoon paprika

½ teaspoon cumin

½ lime (juiced)

1 teaspoon lime zest

1 teaspoon sea salt

1 teaspoon black pepper powder

- Season the pork with garlic, chili powder, cumin, salt, pepper and paprika.
- Add in the lime juice and zest and leave to marinate for 20 minutes.
- Using a foil, line the broiler pan and place the pork chops on it.
- Broil per side for 4-5 minutes.

Per Serving: Calories: 231; Total Fat: 13.2g; Saturated Fat: 5g; Protein: 25.4g; Carbs: 3.4g; Fiber: 0.6g; Sugar: 0.2g;

Pork Cutlets

Serves: 6 / Preparation time: 10 minutes / Cooking time: 12 minutes

These pork cutlets make a wonderful appetizer. With minimal ingredients you can get an appetizer dish for several people, or a simple pork meal for one or two.

12 oz pork (ground)

1 teaspoon fennel seeds

½ teaspoon crushed red pepper

1 teaspoon oregano

Sea salt to taste

Black pepper powder to taste

- Combine all the ingredients in a bowl and mix well.
- Shape portions of the mixture into patties.
- Cook covered in a skillet over medium flame for around 3 minutes per side.

Per Serving: Calories: 84; Total Fat: 2.1g; Saturated Fat: 0.7g; Protein: 14.9g; Carbs: 0.4g; Fiber: 0.3g; Sugar: 0g;

Apple & Cherry Pork

Serves: 4 / Preparation time: 10 minutes / Cooking time: 40 minutes

This fruity pork is a dish that all will relish – both young and old.

1 tablespoon extra-virgin olive oil

2 cups apple (diced)

⅔ cups cherry (pitted)

⅓ cup onion (diced)

⅓ cup celery (diced)

½ cup apple juice

⅛ teaspoon sea salt

⅛ teaspoon Black Pepper

1 ⅓ pounds boneless Pork Loin

- Combine all the ingredients in an Instant Pot and close the lid.
- Select the Meat/Stew function and cook for 40 minutes.
- Do a quick release of pressure.

Per Serving: Calories: 336; Total Fat: 9.2g; Saturated Fat: 2.3g; Protein: 40.3g; Carbs: 23.2g; Fiber: 3.5g; Sugar: 17.3g;

Paleo Beef Recipes

Contents

Beef & Bell Pepper Stir Fry

Serves: 4 / Preparation time: 10 minutes / Cooking time: 8 minutes

This beef and bell pepper stir fry is a colorful dish that tastes just as good as it looks. You can increase the amount of jalapeno peppers if you like it with an extra kick.

1 pound lean sirloin (sliced into strips)

1 yellow bell pepper (sliced thinly)

1 jalapeno pepper (sliced thinly)

1 red bell pepper (sliced thinly)

1 red onion (sliced thinly)

1 tomato (wedged)

1 teaspoon olive oil

4 garlic cloves (minced)

1 teaspoon cumin

1 ½ tablespoon red wine vinegar

¼ teaspoon coconut aminos

1/3 cup cilantro (chopped roughly)

Sea salt to taste

Ground black pepper to taste

- Season the beef with cumin, salt and pepper.

- Heat oil in a skillet and cook the beef in into until browned.

- Mix in the onions, peppers and garlic, stir cooking for 3-4 minutes.

- Mix in the coconut aminos and vinegar, cooking for another 2 minutes.

- Mix in the tomatoes and cilantro, cooking for yet another 2 minutes.

Per Serving: Calories: 260; Total Fat: 8.1g; Saturated Fat: 2.7g; Protein: 36g; Carbs: 9.3g; Fiber: 1.9g; Sugar: 4.8g;

Beef Cutlets

Serves: 4 / Preparation time: 10 minutes / Cooking time: 10 minutes

Serve these beef cutlets as a starter for a dinner party. You can also try this recipe with chicken mince.

1 pound ground beef (95% lean)

1/3 cup onion (diced)

2 garlic cloves (minced)

½ teaspoon sea salt

½ teaspoon ground pepper

½ teaspoon oregano

¼ teaspoon red pepper flakes

1 egg

2 tablespoon fresh basil (chopped)

1 tablespoon tomato paste

1 teaspoon olive oil

- Whisk together the tomato paste and egg.

- Combine the rest of the ingredients except the oil in a bowl, mixing well.

- Add the egg mixture to the bowl and mix well.

- Shape portions of the mixture into patties and brush it with olive oil

- Grill the burgers in a non-stick skillet for 4-5 minutes per side.

Per Serving: Calories: 237; Total Fat: 9g; Saturated Fat: 3.9g; Protein: 36.6g; Carbs: 2.6g; Fiber: 0.6g; Sugar: 1g;

Thai Style Ground Beef Curry

Serves: 4 / Preparation time: 10 minutes / Cooking time: 4 hours 30 minutes

This Thai Style ground beef curry tastes awesome over a hot bowl of cauliflower rice. You can drizzle some lime juice over before serving.

1 pound ground beef (95% lean)

1 leek (thinly sliced)

2 garlic cloves (minced)

1 teaspoon fresh ginger (raw)

1 tablespoon red curry paste

11/2 cups tomato sauce

1 teaspoon lime zest

1 tablespoon coconut aminos

½ cup light coconut milk

- Brown the beef in a skillet and then transfer it to a slow cooker.
- Add the rest of the ingredients to the slow cooker except the coconut milk.
- Leave to cook for 4 hours on low.
- Add the coconut milk, stir and cook for another 15 minutes.

Per Serving: Calories: 386; Total Fat: 15.7g; Saturated Fat: 10.2g; Protein: 40.3g; Carbs: 25.3g; Fiber: 6.3g; Sugar: 16.3g;

Beef Chili

Serves: 6 / Preparation time: 10 minutes / Cooking time: 25 minutes

There are a lot of ingredients to this recipe, but it is actually quite simple to prepare and doesn't take long to cook. This can be an ideal weeknight Paleo meal.

2 pounds ground beef (95% lean)

2 cups chicken broth

1 ½ cups water

½ tablespoon olive oil

½ onion (minced)

2 garlic cloves (minced)

1 green pepper (chopped)

1 poblano pepper (seeded, chopped)

½ teaspoon cumin

2 teaspoon chili powder

½ teaspoon sea salt

½ cup salsa

14 oz fire roasted diced tomatoes (canned)

2 zucchini (chopped)

- Heat the oil in a pot and sauté the garlic, onion and poblano pepper in it for 2-3 minutes.

- Mix in the beef and stir cook until browned.

- Mix in the rest of the ingredients and leave to simmer for around 20 minutes.

Per Serving: Calories: 337; Total Fat: 11g; Saturated Fat: 4.8g; Protein: 50g; Carbs: 9.9g; Fiber: 3.1g; Sugar: 4.2g;

Beef Meatloaf

Serves: 10 / Preparation time: 15 minutes / Cooking time: 1 hour

A classic meatloaf for the Paleo crowd. Some people prefer this dish with an extra tablespoon or two of crushed garlic, so experiment to see how you like it.

1 diced red onion

2 pounds grass-fed ground beef

1 cup almond meal

2 eggs

1 can tomato paste

1 tbsp crushed garlic

1/2 tbsp sea salt

2 tbsp dried basil

1 tsp marjoram

Cracked black pepper to taste

- Mix together all the ingredients in a bowl and then place in a glass loaf pan.
- Bake for an hour in an oven preheated to 350 degrees Fahrenheit/ degrees Celsius.

Per Serving: Calories: 342.4; Total Fat: 25.8g; Saturated Fat: 8.3g; Protein: 20.9g; Carbs: 5.9g; Fiber: 2.1g; Sugar: 2g;

Spinach Beef Balls

Serves: 4 / Preparation time: 15 minutes / Cooking time: 20 minutes

This dish is a great way to make your kids eat their greens. You can replace spinach with other greens like kale or Swiss Chard as well.

1 pound ground beef (99% lean)

10 oz frozen chopped spinach (defrosted, moisture squeezed out)

½ cup onion (minced)

4 garlic cloves (minced)

½ teaspoon sea salt

1 egg

½ teaspoon dried basil

¼ teaspoon oregano

¼ teaspoon pepper

- Combine all the ingredients in a bowl and mix well.
- Shape portions of the mixture into meat balls and place on a greased baking sheet.
- Bake in an oven preheated to 375 degrees Fahrenheit for 19-20 minutes.

Per Serving: Calories: 205; Total Fat: 9.5g; Saturated Fat: 3.4g; Protein: 26.1g; Carbs: 5.1g; Fiber: 2g; Sugar: 1g;

Sour & Spicy Beef

Serves: 6 / Preparation time: 10 minutes / Cooking time: 8 hours

These sour and spicy beef balls are rich in aroma and flavor. You definitely cannot resist these when laid on the table.

2 pounds beef eye round

1 sweet onion (diced)

2 garlic cloves (sliced)

1 red bell pepper (diced)

2 jalapenos

¼ cup chicken broth

1 cup canned crushed tomatoes

1 tablespoon coconut aminos

½ lime (juiced)

½ teaspoon cumin

¼ teaspoon oregano

¼ teaspoon coriander

Black pepper to taste

Sea salt to taste

- Season the beef with salt and pepper and place it in a slow cooker.
- Add the rest of the ingredients to the slow cooker.
- Leave to cook for 8 hours on low.
- Slice and serve.

Per Serving: Calories: 228; Total Fat: 5.6g; Saturated Fat: 2g; Protein: 35.8g; Carbs: 6g; Fiber: 1.2g; Sugar: 2g;

Beef Stew

Serves: 6 / Preparation time: 20 minutes / Cooking time: 6-8 hours

Serve this beef stew piping hot. It will keep you warm on a chilly night.

1 ½ pounds Beef stew meat

1 medium onion

2 tbsp olive oil

1 tbsp basil

1 tsp oregano

1 tsp cilantro leaves

1/4 tsp chili powder

1 clove garlic minced

14 oz can diced tomatoes

1 cup frozen spinach

1 tsp sea salt

1 tsp ground pepper

1 tbsp Italian seasoning

1 tsp thyme

- Mix together all the seasonings and rub it on the beef along with the oil.
- Place the beef in a slow cooker.
- Add the rest of the ingredients to the slow cooker.
- Leave to cook for 6-8 hours on low.

Per Serving: Calories: 397; Total Fat: 25.9g; Saturated Fat: 9.2g; Protein: 24.1g; Carbs: 16.8g; Fiber: 3.2g; Sugar: 1.4g;

Broccoli & Beef Mix

Serves: 4 / Preparation time: 20 minutes / Cooking time: 20 minutes

Broccoli & beef mix is a classic staple of American Chinese food. Some people like to serve this meal on top of some cauliflower fried rice.

1 pound sirloin steak (sliced)

2 teaspoon black pepper

2 tablespoon coconut oil

2 teaspoon ginger (grated)

½ cup chicken broth

2 garlic cloves (minced)

¼ teaspoon sea salt

2 cups broccoli (florets)

2 cups carrots (sliced thin)

1 green onion (sliced thin)

½ teaspoon red pepper flakes

- Heat a tablespoon of coconut oil in a skillet and sauté the garlic in it.

- Season the beef with salt and brown it in the skillet. Transfer into a platter.

- Heat the other tablespoon of coconut oil in a skillet and stir fry the carrots and broccoli in it.

- Add the rest of the ingredients to the skillet except the beef and greens onions and cook until the broccoli is tender.

- Add the beef and green onions to the skillet and cook until reheated.

Per Serving: Calories: 331.1; Total Fat: 16.7g; Saturated Fat: 9.5g; Protein: 36.8g; Carbs: 8.5g; Fiber: 3.2g; Sugar: 1.9g;

Beef in Marinara Sauce

Serves: 6 / Preparation time: 15 minutes / Cooking time: 45 minutes

This beef in marinara sauce is a perfect recipe to make in batches. You can freeze portions and enjoy them later in the week for a quick and easy microwavable lunch.

1 pound lean ground beef (browned)

1 tablespoon olive oil

1 cup baby carrots (diced)

1 cup mushrooms (diced)

5 garlic cloves (chopped)

2 teaspoon dried basil

2 teaspoon dried oregano

2 bay leaves (dried)

1 can diced tomatoes (drained)

1 yellow pepper (diced)

2 teaspoon dried thyme

Ground black pepper to taste

1 can crushed tomatoes

1 onion (chopped)

- Heat oil in a pot and sauté the onion and garlic in it.
- Add in the herbs, bay leaves, yellow pepper, carrots, mushrooms and black pepper and simmer until the veggies soften.
- Pour in the tomato cans and bring to boil, stirring often.
- Reduce the flame and mix in the browned beef, leaving to simmer for 40-45 minutes.

Per Serving: Calories: 266.1; Total Fat: 18.3g; Saturated Fat: 6.7g; Protein: 15.4g; Carbs: 10.5g; Fiber: 2.7g; Sugar: 2.9g;

Paleo Lamb Recipes

Contents

Lamb Curry

Serves: 6 / Preparation time: 10 minutes / Cooking time: 25 minutes

This lamb curry is so simple to make. Cooking it in a pressure cooker or an Instant Pot brings its rich flavor out really well.

1 ½ pounds lamb stew meat (diced)

½ cup coconut milk

4 garlic cloves (minced)

½ lime (juiced)

1 ½ tablespoon garam masala

3 teaspoon ginger (grated)

1 tablespoon olive oil

1 (14 ounce) can diced tomatoes

¾ teaspoon turmeric

1 onion (cubed)

3 carrots (sliced)

1 zucchini (cubed)

¼ teaspoon sea salt

Ground black pepper to taste

- Combine the lamb, garlic, ginger, lime juice, coconut milk, salt and pepper and marinate for 8 hours.
- Place the lamb in a pressure cooker or an Instant Pot and add the rest of the ingredients to it except the zucchini.
- Cook for 20 minutes on high pressure (select the "Manual" function on an Instant Pot)
- Do a natural release of pressure.
- Add the zucchini to the Instant Pot and set it to the Sauté function (or sauté in a pan lightly drizzled with olive oil on medium heat) and cook for 5 minutes.

Per Serving: Calories: 319; Total Fat: 15.7g; Saturated Fat: 7.6g; Protein: 33.9g; Carbs: 10.8g; Fiber: 2.9g; Sugar: 5.4g;

Spiced Lamb Leg

Serves: 6 / Preparation time: 10 minutes / Cooking time: 4 hours

This spiced lamb leg is so moist and tempting that it's surely going to boost your appetite.

70 ½ oz leg of lamb

2 cups water

1 tablespoon ground coriander

1 tablespoon olive oil

1 teaspoon sea salt

1 tablespoon ground cumin

- Pour the water into a roasting pan.

- Rub the spices and salt onto the lamb and place the lamb in the roasting pan.

- Drizzle the oil over the lamb.

- Cook in an oven preheated to 285 degrees Fahrenheit/140 degrees Celsius for 4 hours.

Per Serving: Calories: 643; Total Fat: 27g; Saturated Fat: 9.1g; Protein: 93.7g; Carbs: 0.5g; Fiber: 0.1g; Sugar: 0g;

Lamb Rib Chops

Serves: 8 / Preparation time: 10 minutes / Cooking time: 30 minutes

These lamb rib chops are indeed a delicacy. It will leave you totally amazed by the flavor that simple ingredients can have.

48 oz Lamb Ribs

2 tablespoon olive oil

1 teaspoon sea salt

½ teaspoon black pepper

2 tablespoon parsley

2 teaspoon rosemary

2 tablespoon peppermint

4 tablespoon Dijon Mustard

2 garlic cloves

- Season the lamb with salt and pepper.

- Preheat an oven to 425 degrees Fahrenheit/204.4 degrees Celsius.

- Mix the mint, parsley, rosemary, and garlic.

- Place the lamb in a roasting pan along with the oil.

- Brush the mustard on both sides of the lamb and spread the herb mixture over.

- Roast for 25-30 minutes.

- Leave to cool and then slice.

Per Serving: Calories: 1179; Total Fat: 63.9g; Saturated Fat: 21.5g; Protein: 138.5g; Carbs: 1.1g; Fiber: 0.6g; Sugar: 0.1g;

Lamb Shanks

Serves: 3 / Preparation time: 10 minutes / Cooking time: 8 hours

You can prepare these lamb shanks with great ease. All you need is a few basic ingredients and a slow cooker.

3 pounds lamb shanks

8 garlic cloves

3 tablespoon fresh rosemary (chopped)

2 carrots (chopped)

1 onion (sliced)

Sea salt to taste

Ground black pepper to taste

- Season the lamb with salt and pepper.

- Place the lamb in a slow cooker with the remaining ingredients.

- Pour sufficient water to just cover the meat.

- Cook for 7-8 hours on low.

Per Serving: Calories: 889; Total Fat: 33.4g; Saturated Fat: 11.9g; Protein: 128.7g; Carbs: 10.4g; Fiber: 2.2g; Sugar: 3.6g;

Lamb Stew

Serves: 4 / Preparation time: 10 minutes / Cooking time: 40 minutes

This lamb stew is packed with flavor. Serve it to guests and you will be delighted to see the appreciation you receive as a master Paleo chef.

1 pound ground lamb

2 tablespoon olive oil

½ cup celery (chopped)

½ cup red bell pepper (chopped)

1 (6 oz) can tomato paste

2 cups packed kale leaves (chopped)

1 cup carrots (chopped)

½ teaspoon ground cumin

1 teaspoon bay leaf

2 cups chicken stock

2 garlic cloves

½ teaspoon sea salt

½ teaspoon ground black pepper

- Heat oil in a skillet and sauté the carrots, bell pepper, onion and celery in it for 7-8 minutes.

- Mix in the lamb and cook for 6-7 minutes.

- Add in the rest of the ingredients except the stock and cook for 3-4 minutes.

- Pour in the stock and simmer for 20-25 minutes.

Per Serving: Calories: 365; Total Fat: 16g; Saturated Fat: 4.1g; Protein: 36.1g; Carbs: 20.5g; Fiber: 4.3g; Sugar: 9.4g;

Lamb Burgers

Serves: 4 / Preparation time: 10 minutes / Cooking time: 16 minutes

These lamb burgers are a Greek inspired dish. Simple to make and a nice change from traditional hamburgers.

1 1/3 pound ground lamb

1/3 cup fresh mint (chopped)

½ teaspoon sea salt

3 tablespoons olive oil

2 teaspoon paprika

¾ teaspoon ground cinnamon

- Combine all the ingredients in a bowl, leaving aside 1 ½ tbsp oil.

- Mix well and shape portions of the mixture into burgers.

- Heat the remaining oil in a skillet and cook the burgers in it for 4 minutes on each side.

Per Serving: Calories: 379; Total Fat: 21.8g; Saturated Fat: 5.5g; Protein: 42.9g; Carbs: 1.6g; Fiber: 1.1g; Sugar: 0.1g;

Grilled Lamb Chops

Serves: 4 / Preparation time: 10 minutes / Cooking time: 16 minutes

Lamb chops when grilled taste great. You can try these lamb chops at your next barbeque cookout.

8 lamb leg chops

2 garlic cloves (minced)

2/3 cup extra-virgin olive oil

1 tablespoon fresh oregano (minced)

Sea salt to taste

Ground black pepper to taste

1 tablespoon fresh lemon zest

- Season the chops with salt and pepper.

- Mix together the remaining ingredients and brush the chops with the mixture on both sides.

- Preheat the grill and cook the chops on it for 4-5 minutes per side.

Per Serving: Calories: 631; Total Fat: 49.9g; Saturated Fat: 10.8g; Protein: 44.1g; Carbs: 1.5g; Fiber: 0.6g; Sugar: 0.2g;

Herbed Lamb Chops

Serves: 4 / Preparation time: 10 minutes / Cooking time: 16 minutes

Unlike most marinades that take a long time to release the flavor, this herb marinade gets into the chops really quick.

6 lamb loin chops

4 garlic cloves (minced)

2 tablespoon extra-virgin olive oil

1 tablespoon fresh rosemary (crushed)

2 teaspoon sea salt

1 tablespoon fresh thyme (crushed)

- Combine the garlic, salt, rosemary, thyme and half the oil, mix well and marinate the chops in it for 30 minutes.

- Heat the rest of the oil in a skillet and brown the chops in it for 3 minutes each side.

- Place the skillet in an oven preheated to 400 degrees Fahrenheit/ 204 degrees Celsius for 10 minutes.

Per Serving: Calories: 456; Total Fat: 39.2g; Saturated Fat: 16.1g; Protein: 25.3g; Carbs: 2g; Fiber: 0.7g; Sugar: 0g;

Barbacoa Lamb

Serves: 12 / Preparation time: 10 minutes / Cooking time: 6 hours

You can use this barbacoa lamb as a stuffing in lettuce wraps. If you like it with some extra spice you could add some chili flakes as desired.

5 ½ lbs leg of lamb (boneless)

2 tablespoon Himalayan salt

1 tablespoon ground cumin

1 teaspoon chipotle powder

2 tablespoon smoked paprika

1 cup water

¼ cup dried mustard

1 tablespoon dried oregano

- Rub the lamb with the mustard.

- Mix together the rest of the ingredients in a bowl except the water and spread the mixture over the lamb.

- Place the lamb in a slow cooker and pour in the water.

- Cook for 6 hours on high.

- Shred the lamb and mix in a cup of cooking liquid.

Per Serving: Calories: 492; Total Fat: 35.8g; Saturated Fat: 14.8g; Protein: 37.5g; Carbs: 1.2g; Fiber: 0.5g;

Lamb Shanks with Mushroom Gravy

Serves: 4 / Preparation time: 15 minutes / Cooking time: 6-8 hours

When cooked these lamb shanks become amazingly tender and when topped with the luscious mushroom gravy, these taste heavenly.

1 ½ pounds lamb shanks	16 oz button mushrooms (diced)
4 carrots (peeled, diced)	4 tablespoon tomato paste
8 tablespoon shallots (peeled, diced)	2 tablespoon bay leaves (crumbled)
4 tablespoon olive oil	3 cups chicken broth
4 garlic cloves (minced)	Sea salt to taste
4 celery stalks (diced)	Ground black pepper to taste

- Heat half the oil in a skillet and brown the lamb shanks in it on both sides in batches.

- Mix together the rest of the ingredients in a crock pot and place the lamb shanks in it.

- Cook for 6-8 hours on low.

Per Serving: Calories: 552; Total Fat: 28.1g; Saturated Fat: 6.8g; Protein: 57g; Carbs: 19g; Fiber: 3.9g; Sugar: 7.7g;

Paleo Seafood Recipes

Contents

Mushroom Seafood Skillet

Serves: 6 / Preparation time: 10 minutes / Cooking time: 15 minutes

This mushroom seafood skillet is a one pot meal. It is easy to make and easy to clean up. The perfect weekday Paleo meal!

4 oz shelled shrimps (raw)

4 oz smoked salmon (chopped into strips)

1 cup mushrooms (sliced)

4 oz uncured beef bacon slices (chopped)

½ cup coconut cream

Pinch of Sea salt

Ground pepper to taste

- Heat a cast iron skillet and sauté the bacon in it until cooked.

- Add the mushrooms and stir cook for 5 more minutes.

- Add in the salmon, stir cooking for 2 minutes.

- Increase the flame, and add the shrimps, stir cooking for 2 more minutes.

- Season with salt and pepper and mix in the cream, cooking for 1 additional minute.

Per Serving: Calories: 272; Total Fat: 11.7g; Saturated Fat: 7g; Protein: 31.2g; Carbs: 2.3g; Fiber: 0.9g; Sugar: 1.3g;

Hazelnut Smothered Cod Fillets

Serves: 2 / Preparation time: 10 minutes / Cooking time: 20 minutes

You can even prepare this dish with sea bass filets or haddock filets. The hazelnut crunchy coating gives the fish a rich flavor.

2 cod filets

2 tablespoon almond butter

½ teaspoon garlic powder

½ teaspoon sea salt

½ teaspoon pepper

1/8 teaspoon cayenne

2 lemon slices

1 tablespoon parsley (chopped)

1/3 cup toasted hazelnuts

Ground pepper to taste

- Mix together the nuts, garlic powder, salt, pepper and cayenne pepper in a food processor and pulse until crumbed.

- Add the almond butter to a roast pan and place in an oven preheated to 425 degrees Fahrenheit/218.3 degrees Celsius.

- Once the butter has melted, place the filets in the roast pan skin-side down and spoon the butter over it.

- Sprinkle the hazelnut crumb mixture over the filets, pressing it lightly

- Squeeze the lemon juice over the fish and place the slices over.

- Cook for 12-15 minutes in the oven.

- Serve topped with some of the butter and parsley.

Per Serving: Calories: 376; Total Fat: 17.7g; Saturated Fat: 1.3g; Protein: 25.6g; Carbs: 32.1g; Fiber: 2.1g; Sugar: 18.5g;

Roasted Salmon with Leeks

Serves: 2 / Preparation time: 10 minutes / Cooking time: 20 minutes

This roasted salmon dish is so tender and moist that the fish just melts in your mouth. Salmon is one of the healthiest fishes you can eat, so don't feel guilty about enjoying this recipe regularly.

12 oz wild pink salmon filets

2 tablespoon oil

1 tablespoon fresh dill (chopped)

1 tablespoon lemon zest

¼ teaspoon pepper

½ teaspoon sea salt

2 leeks (white and green portions, chopped)

- Mix together the salt, pepper, dill and zest.

- Brush the fish with some oil and spread the dill mixture over the filets.

- Place the fish in a roasting pan along with the rest of the oil and cook for around 15 minutes in an oven preheated to 425 degrees Fahrenheit/218.3 degrees Celsius, adding the leeks during the last 3 minutes of cooking.

Per Serving: Calories: 337; Total Fat: 15.9g; Saturated Fat: 2g; Protein: 37.7g; Carbs: 15.6g; Fiber: 1.9g; Sugar: 5.3g;

Herbed Mahi Mahi Fillets

Serves: 2 / Preparation time: 10 minutes / Cooking time: 15 minutes

Serve these Mahi Mahi fillets with a helping of roasted veggies for a complete lunch. You can also use catfish for this recipe.

2 (4 oz) mahi mahi filets	¼ teaspoon sea salt
1 teaspoon olive oil	½ teaspoon pepper
1 ½ teaspoon paprika	1 teaspoon lemon juice
½ teaspoon dried basil	1/8 teaspoon cayenne pepper
½ teaspoon dried tarragon	

- Mix together the lemon juice and oil and brush it over the filets.

- Mix together the rest of the ingredients and rub it on both sides of the fish.

- Place the filets in a roast pan.

- Cook in an oven preheated to 350 degrees Fahrenheit/176.6 degrees Celsius for 10-15 minutes.

Per Serving: Calories: 427; Total Fat: 6.6g; Saturated Fat: 0.4g; Protein: 84.4g; Carbs: 5.4g; Fiber: 0.8g; Sugar: 0.2g;

Coconut Buttered Haddock

Serves: 4 / Preparation time: 10 minutes / Cooking time: 30 minutes

Coconut buttered haddock is a refreshing, tropical take on a popular fish filet.

1 lb haddock filets (cut into pieces of serving size)

¼ cup coconut butter (melted)

2 tablespoon water

½ teaspoon garlic salt

½ teaspoon dill weed

¼ cup green onions (chopped)

¼ cup lemon juice

- Mix together the lemon juice, coconut butter, dill, garlic salt, water and onions in a roasting pan.

- Add the fish filets and to the pan and coat on both sides.

- Refrigerate covered on both sides.

- Remove from the refrigerator and loosely cover with a foil.

- Cook in an oven preheated to 350 degrees Fahrenheit/176.6 degrees Celsius for 25-30 minutes.

Per Serving: Calories: 250; Total Fat: 11.5g; Saturated Fat: 9.6g; Protein: 29.3g; Carbs: 5.1g; Fiber: 2.3g; Sugar: 1.6g;

Roasted Shrimps with Veggies

Serves: 24 / Preparation time: 10 minutes / Cooking time: 30 minutes

Shrimps are a favorite type of seafood because they are so juicy and tasty. With all of the vegetables in this recipe, this is a supremely healthy Paleo meal.

1 ½ lbs shrimps (peeled, deveined)

2 zucchini (sliced into half-moon shapes)

1 yellow summer squash (sliced into half-moon shapes)

1 cup cherry tomatoes (halved)

1 yellow bell pepper (sliced)

3 tablespoon balsamic vinegar

¼ cup extra virgin olive oil

1 red onion (chunked)

1 teaspoon black pepper

3 tablespoon lemon juice

¼ cup basil leaves (chopped)

Sea salt to taste

- Toss together all the veggies with salt, pepper, ¼ cup oil, lemon juice and 2 tablespoon vinegar in a roasting pan.
- Place the shrimps and veggies in a roasting pan lined with parchment paper.
- Roast for 20-25 minutes in an oven preheated to 375 degrees Fahrenheit/190.5 degrees Celsius on the center rack.
- Toss together the shrimps with the rest of the ingredients except the basil and leave to marinate as the veggies are cooking.
- Add the shrimps to the roast pan with the veggies along with the marinade and spread.
- Turn the oven to broil and place the pan on the top rack of the oven, for around 5 minutes.
- Serve garnished with basil.

Per Serving: Calories: 58; Total Fat: 2.7g; Saturated Fat: 0.3g; Protein: 6g; Carbs: 2.1g; Fiber: 0.6g; Sugar: 1.1g;

Lobster & Asparagus Mix

Serves: 2 / Preparation time: 10 minutes / Cooking time: 20 minutes

This lobster and asparagus salad is a light meal that is easy to make. Serve it as a side, or on its own.

8 oz lobster (cooked, chopped)

3 ½ cups asparagus (chopped, steamed)

2 tablespoon lemon juice

4 teaspoons extra-virgin olive oil

¼ teaspoon sea salt

Black pepper to taste

½ cup cherry tomatoes (halved)

1 basil leaf (chopped)

2 tablespoon red onion (diced)

- Clean the lobster and cook it in a skillet.

- Whisk together the lemon juice, salt, pepper and oil in a bowl.

- Toss together the rest of the ingredients including the cooked lobster in a bowl.

- Pour the lemon mixture over and toss again.

Per Serving: Calories: 244; Total Fat: 10.8g; Saturated Fat: 1.8g; Protein: 27.4g; Carbs: 12.1g; Fiber: 5.7g; Sugar: 6.3g;

Salmon & Asparagus Bake

Serves: 4 / Preparation time: 10 minutes / Cooking time: 25 minutes

Baking Salmon is much healthier than frying it and serving it with some asparagus adds to the nutritional value of the meal.

1 ½ pounds coho salmon (skin removed)

2 tablespoon olive oil

2 lemons

4 dill sprigs

1 ½ lb asparagus

Sea Salt to taste

Ground black pepper to taste

- Divide the asparagus among 4 foil pieces and spray it with a cooking spray.

- Arrange the salmon on top, seasoning it with salt and pepper.

- Add the oil over each piece of salmon and then place some lemon slices and dill sprig.

- Fold the foil packets to close it and then place on a baking sheet.

- Bake in an oven preheated to 400 degrees Fahrenheit for 25 minutes.

Per Serving: Calories: 328; Total Fat: 17.8g; Saturated Fat: 2.6g; Protein: 37.1g; Carbs: 9.4g; Fiber: 4.4g; Sugar: 3.9g;

Shrimps & Broccoli Stir Fry

Serves: 4 / Preparation time: 10 minutes / Cooking time: 10 minutes

A quick stir fry dish to prepare if you are a lover of shrimps. The lemon pepper spice adds a unique flavor to the dish.

4 teaspoon olive oil

2 lbs raw shrimps

3 teaspoon lemon pepper spice

¼ cup lemon juice

4 cups broccoli florets

2 tablespoon parsley (chopped)

Sea salt to taste

Ground black pepper to taste

- Heat olive oil in a skillet and cook the shrimps in it for a minute.

- Add the lemon pepper spice and the lemon juice and cook for 3-4 minutes, stirring occasionally until the shrimps are cooked. Remove and place aside.

- Add the broccoli to the skillet and cook covered for another 3-4 minutes.

- Mix the shrimps back in and toss.

- Mix in the parsley and season with salt and pepper.

Per Serving: Calories: 345; Total Fat: 9g; Saturated Fat: 2g; Protein: 54.4g; Carbs: 9.9g; Fiber: 2.5g; Sugar: 1.9g;

Salmon Crusted with Coconut

Serves: 5 / Preparation time: 10 minutes / Cooking time: 12 minutes

Instead of using regular old breadcrumbs, you can add shredded coconut to stay Paleo compliant. It improves the taste as well.

30 oz salmon (skinless, boneless)

1/3 cup unsweetened coconut (shredded)

2 egg whites (whisked)

Sea salt to taste

Ground black pepper to taste

- Season the salmon with salt and pepper.

- Dip the salmon into the egg and then dredge it with the coconut.

- Arrange the fish on a wire rack placed over a baking dish.

- Bake for 10-12 minutes in an oven preheated to 400 degrees Fahrenheit.

Per Serving: Calories: 309; Total Fat: 17.8g; Saturated Fat: 8g; Protein: 35.3g; Carbs: 3g; Fiber: 2g; Sugar: 0.9g;

Paleo Side Dish & Sauce Recipes

Contents

Stuffed Mushrooms

Serves: 1 / Preparation time: 15 minutes / Cooking time: 20 minutes

One of the great things about Portobello mushrooms is how versatile they are. You can stuff them with a variety of meats and veggies. Don't be afraid to experiment with this recipe by substituting ingredients for your favorite veggies and meats.

1 Portobello mushroom

¼ cup baby zucchini (chopped)

4 oz extra-lean ground beef

1/3 cup fresh mushrooms

¼ cup tomato paste

½ tablespoon extra-virgin olive oil

¼ cup canned tomatoes

½ cup water

¼ cup onions (chopped)

- Heat oil in a pan and sauté the onions in it.
- Add the beef and brown it.
- Mix in the remaining ingredients except the Portobello mushroom and cook completely.
- Stuff the mixture into the Portobello mushroom and place on a greased baking dish.
- Bake for 15 minutes in an oven preheated to 350 degrees Fahrenheit/ 175 degrees Celsius.

Per Serving: Calories: 451.3; Total Fat: 27.2g; Saturated Fat: 8.9g; Protein: 29.7g; Carbs: 27.1g; Fiber: 6.9g; Sugar: 12.8g;

Veggie Stir Fry

Serves: 1 / Preparation time: 10 minutes / Cooking time: 10 minutes

This amazing vegetable stir fry is power packed with nutrients. You can serve this aside your favorite meat dishes. Add some sesame seeds and purple cabbage for a bit of color that will make the presentation "pop".

6 asparagus spears (chopped)

1 baby zucchini (chopped

1 teaspoon garlic powder

3 mushrooms (chopped)

1 teaspoon onion powder

½ tablespoon extra-virgin olive oil

1 onion (chopped)

1 teaspoon ground black pepper

1 tablespoon water

- Heat oil in a pan and sauté the onions along with the spices in it until caramelized.
- Add the veggies and leave to cook until lightly browned.
- Add the water and simmer covered for 5 minutes.

Per Serving: Calories: 163.3; Total Fat: 7.8g; Saturated Fat: 1.1g; Protein: 6.8g; Carbs: 21.5g; Fiber: 5.9g; Sugar: 2.7g;

Hollandaise Sauce

Serves: 4 / Preparation time: 5 minutes / Cooking time: 10 minutes

Warm Hollandaise sauce can be served over seafood, eggs and veggies.

3 egg yolks

2 teaspoon water

1/8 teaspoon paprika

¼ teaspoon sea salt

2 teaspoon apple cider vinegar

¼ cup olive oil

- Whisk together the water and egg in a saucepan over low flame until creamy, adding a little water at a time if required.

- Gently pour in the oil while whisking.

- Mix in the vinegar, paprika and sea salt and whisk again.

Per Serving: Calories: 165.4; Total Fat: 17.2g; Saturated Fat: 4.9g; Protein: 2.1g; Carbs: 0.5g; Fiber: 0g; Sugar: 0.1g;

Mustard Sauce

Serves: 24 / Preparation time: 1 minutes / Cooking time: 0 minutes

You can serve this mustard sauce as a dip for appetizers or you can also serve it as a spread on some fresh veggies.

1 egg

1 cup olive oil

Sea salt to taste

1 teaspoon Dijon mustard

Ground black pepper to taste

1 tablespoon lemon juice

- Combine the lemon juice, Dijon mustard and egg in a container and pour in the oil.
- Leave aside for a few seconds and then using an immersion blender, blend for 25 seconds without moving the blender.

Per Serving: Calories: 82.8; Total Fat: 9.2g; Saturated Fat: 1.3g; Protein: 0.3g; Carbs: 0g; Fiber: 0g; Sugar: 0g;

Barbeque Sauce

Serves: 4 / Preparation time: 5 minute / Cooking time: 15 minutes

This barbeque sauce has a tangy flavor that goes perfectly well with all types of meats. The natural sugar content is a little high however, so enjoy in moderation.

2 tablespoon apple cider vinegar

12 oz tomato paste

Sea salt to taste

2 tablespoon maple syrup

1 teaspoon garlic powder

1 tablespoon fish sauce

Ground black pepper to taste

2 tablespoon molasses

- Mix together all the ingredients in a saucepan.
- Leave to simmer on low flame for around 15 minutes.

Per Serving: Calories: 129.7; Total Fat: 0.4g; Saturated Fat: 0.1g; Protein: 4g; Carbs: 31.4g; Fiber: 3.9g; Sugar: 22.5g;

Green Beans with Bacon

Serves: 8 / Preparation time: 5 minute / Cooking time: 15-20 minutes

These green beans with bacon make for a tasty side dish on your dinner table. Who can resist the taste of bacon?

12 cups green beans

¾ cups bacon (diced)

4 teaspoon garlic (minced)

1 tablespoon sea salt

2 teaspoon ground black pepper

- Toss together all the ingredients in a baking pan lined with parchment paper.
- Bake for 15-20 minutes in an oven preheated to 350 degrees Fahrenheit/ 175 degrees Celsius.

Per Serving: Calories: 105; Total Fat: 4g; Protein: 6g; Carbs: 13g; Fiber: 6g; Sugar: 2g;

Roasted Cauliflower

Serves: 4 / Preparation time: 5 minute / Cooking time: 15-20 minutes

The savory taste of this dish will surely amaze you, because all you are using is basic seasonings available in your pantry.

3 cups cauliflower (diced)

2 teaspoon coconut oil

2 teaspoon garlic (minced)

1/8 teaspoon sea salt

½ teaspoon chilli powder

1/8 teaspoon ground black pepper

- Toss together all the ingredients in a baking pan lined with parchment paper.
- Bake for 25-30 minutes in an oven preheated to 425degrees Fahrenheit/ 218 degrees Celsius.

Per Serving: Calories: 42; Total Fat: 2g; Saturated Fat: 2g; Protein: 2g; Carbs: 5g; Fiber: 2g; Sugar: 2g;

Spaghetti Sauce

Serves: 15 / Preparation time: 15 minute / Cooking time: 50 minutes

This homemade spaghetti sauce is healthy and packed with Italian flavors. You can serve it over spiralized veggies.

1 1/2 pounds ground beef

2 (29 oz) cans tomato puree

2 teaspoons basil

1 teaspoon thyme

2 (14.5 oz) cans diced tomatoes, with juice

3 teaspoons minced garlic

2 teaspoons lemon juice

2 tablespoons olive oil

2 teaspoons oregano

1 teaspoon crushed red pepper

2 teaspoons sea salt

- Heat a skillet and brown the beef and garlic in it. Drain.
- Mix in the rest of the ingredients and bring to boil, stirring frequently.
- Reduce the flame and simmer uncovered for 45 minutes.

Per Serving: Calories: 163; Total Fat: 11g; Saturated Fat: 3g; Protein: 9g; Carbs: 7g; Fiber: 1g; Sugar: 4g;

Herb Flavored Veggies

Serves: 4 / Preparation time: 10 minutes / Cooking time: 10 minutes

The perfect combination of herbs makes this healthy veggie dish a favorite side. Serve it with steaks or pork chops.

1 red onion (chopped)

1 zucchini (sliced)

1 yellow bell pepper (chopped)

2 tomatoes (chopped)

1 teaspoon lemon pepper

1 teaspoon dried parsley

½ teaspoon garlic powder

1 teaspoon dried oregano

1 tablespoon olive oil

Sea salt to taste

- Heat oil in a skillet over medium flame and sauté the onions in it.
- Mix in the zucchini and cook for another 3 minutes.
- Mix in the tomatoes and bell pepper and cook until tender.
- Add the rest of the ingredients and mix well.

Per Serving: Calories: 72; Total Fat: 3.9g; Saturated Fat: 0.6g; Protein: 1.8g; Carbs: 9.4g; Fiber: 2.5g; Sugar: 5.3g;

Grilled Broccoli

Serves: 4 / Preparation time: 10 minutes / Cooking time: 10 minutes

A simple broccoli side that is jazzed up by garlic powder and red pepper flakes. No, it isn't the most interesting side in the world, but it is healthy, simple, and the taste is pleasing even to those who don't like broccoli.

4 cups broccoli florets

½ tablespoon garlic powder

½ tablespoon sea salt

½ tablespoon ground black pepper

¼ teaspoon red pepper flakes

2 tablespoon olive oil

- Toss together all the ingredients and spread it in a greased baking pan.
- Grill for 8-10 minutes until crispy tender.

Per Serving: Calories: 97; Total Fat: 7.4g; Saturated Fat: 1g; Protein: 2.8g; Carbs: 7.4g; Fiber: 2.7g; Sugar: 1.8g;

Paleo Dessert & Snack Recipes

Contents

Fruit Salad

Serves: 10 / Preparation time: 15 minutes / Cooking time: 0 minutes

A simple mix of fruit that is so refreshing when served chilled. You can always substitute fruits for more variety if you are eating fruit salad often.

2 bananas (peeled, diced)

1 cup pineapple (peeled, diced, juice reserved)

2 mangoes (peeled, diced)

1 papaya (peeled, diced)

¼ cup coconut (freshly grated)

- Toss together the pineapple along with the juice, mangoes and papaya in a bowl and refrigerate covered.
- Prior to serving toss in the banana.
- Serve garnished with coconut.

Per Serving: Calories: 90; Total Fat: 1.1g; Saturated Fat: 0.7g; Protein: 1.1g; Carbs: 21.3g; Fiber: 2.6g; Sugar: 16.3g;

Rhubarb & Strawberry Compote

Serves: 11 / Preparation time: 10 minutes / Cooking time: 15 minutes

A simple Paleo dessert that is sweet, but still healthy.

1 lb strawberries (washed, dried, hulled)

2 tablespoon honey

1 lb rhubarb (leaves discarded, stalks cleaned)

1 tablespoon water

- Chop the strawberries into quarters and the rhubarb into slices of 2 inches
- Combine all the ingredients in a pan and bring to boil.
- Reduce the flame and simmer for 15 minutes until the fruit softens.

Per Serving: Calories: 33; Total Fat: 0.2 g; Saturated Fat: 0g; Protein: 0.7g; Carbs: 8.2g; Fiber: 1.6g; Sugar: 5.6g;

Fruity Pops

Serves: 4 / Preparation time: 10 minutes / Cooking time: 0 minutes

You can have fun replacing the strawberries with black berries or blueberries to get varied colored fruity pops. A great recipe to make with kids.

1/3 cup kiwi (diced)

1/3 cup strawberries (diced)

1/3 cup pineapple (diced)

1/3 cup watermelon (diced)

¼ cup pineapple juice (fresh)

- Toss together the fruit in a bowl.
- Divide the fruit into four 5 oz cups and add a tablespoon of pineapple juice into each.
- Insert ice cream sticks into each cup.
- Freeze for a couple of hours till firm.

Per Serving: Calories: 32; Total Fat: 0.2 g; Saturated Fat: 0g; Protein: 0.5g; Carbs: 7.8g; Fiber: 1g; Sugar: 5.6g;

Zucchini & Turnip Burgers

Serves: 2 / Preparation time: 10 minutes / Cooking time: 6 minutes

These zucchini and turnip burgers are healthy snacks. Your children will love these without knowing the ingredients.

1 cup turnip (grated)

½ cup zucchini (grated, moisture removed)

1 onion (grated)

1 teaspoon garlic powder

2 tablespoon olive oil

Sea salt and ground black pepper to taste

1 tablespoon fresh thyme

- Mix together all the ingredients in a bowl except the oil.

- Heat some oil in a pan and add the small portions of the mixture into the pan.

- Cook until golden brown and then flip, cooking on the other side.

- Repeat with the remaining mixture.

Per Serving: Calories: 176; Total Fat: 14.4g; Saturated Fat: 2.1g; Protein: 2.4g; Carbs: 12g; Fiber: 3.1g; Sugar: 6g;

Cinnamon Pumpkin Fudge

Serves: 25 / Preparation time: 10 minutes / Cooking time: 0 minutes

This cinnamon pumpkin fudge has a wonderful flavor of coconut and spice. You definitely cannot resist these once you taste them.

1 cup pumpkin puree

1 teaspoon ground cinnamon

¼ teaspoon ground nutmeg

1 ¾ cups melted coconut butter (warm)

1 tablespoon coconut oil

- Mix together the coconut butter, pumpkin and spices and then whisk in the coconut oil.

- Spread the mixture into baking pan lined with foil and cover with wax paper, pressing the mixture evenly.

- Discard the wax paper and refrigerate for 2 hours.

- Chop into squares.

Per Serving: Calories: 132; Total Fat: 11.8g; Saturated Fat: 11.1g; Protein: 1.2 g; Carbs: 5.4g; Fiber: 2.6g; Sugar: 1.5g;

Crab & Salmon Burgers

Serves: 14 / Preparation time: 10 minutes / Cooking time: 40 minutes

Serve these crab and salmon burgers to seafood lovers and it will surely bring a smile to their face.

2 eggs

1 lb canned crab meat

14.75 oz canned salmon

2 tablespoon coconut flour

½ jalapeno (diced)

½ onion (diced)

1 cup green onion (chopped)

1 teaspoon Old Bay seasoning

- Whisk together the eggs and the seasoning.

- Mix in the remaining ingredients and then shape portion of the mixture into burgers.

- Heat the oil in a skillet and cook the burgers in it on both sides.

Per Serving: Calories: 95.5; Total Fat: 3.1g; Saturated Fat: 0.7g; Protein: 14.3 g; Carbs: 2.2g; Fiber: 0.7g; Sugar: 0.2g;

Chocolate Brownie

Serves: 2 / Preparation time: 10 minutes / Cooking time: 1 minute

This is an ideal snack or dessert when you are craving some chocolate and it can be whipped up in no time at all.

1 egg

2 tablespoon cocoa powder (unsweetened)

1 tablespoon coconut oil

2 tablespoon almond butter

½ teaspoon vanilla extract

½ teaspoon Paleo baking powder

1 tablespoon honey

Dashes of cinnamon

- Mix together all the ingredients well and place in a mug.

- Microwave the mixture for a minute.

Per Serving: Calories: 237; Total Fat: 18.7g; Saturated Fat: 7.7g; Protein: 7.2 g; Carbs: 15.6g; Fiber: 3.3g; Sugar: 9.7g;

Zucchini Crisps

Serves: 4 / Preparation time: 10 minutes / Cooking time: 12 minute

These zucchini crisps are so addictive. Sit with a bowl of them as you watch your next movie and enjoy the flavor and dispensing with the feeling of guilt associated with eating potato chips. These chips are healthy and Paleo.

1 large zucchini (sliced into circles)

1 cup almond flour

1 egg (beaten)

1 teaspoon fine-grain sea salt

1 teaspoon garlic powder

1 teaspoon thyme

¼ teaspoon ground black pepper

- Combine all the dry ingredients in a bowl.

- Dip the zucchini, first in egg and then dredge it in the flour mixture.

- Place the zucchini circles in a baking dish lined with parchment paper.

- Bake for 6 minutes in an oven preheated to 450 degrees Fahrenheit/ 230 degrees Celsius.

- Flip and bake for another 6 minutes.

Per Serving: Calories: 72; Total Fat: 4.8g; Saturated Fat: 0.6g; Protein: 4 g; Carbs: 5g; Fiber: 1.8g; Sugar: 1.9g;

Coconut Chicken Fingers

Serves: 4 / Preparation time: 10 minutes / Cooking time: 25 minute

Chicken Fingers are a delight to the taste buds. You can make these as a filling snack when you are craving some meat.

1 pound chicken tenders

½ teaspoon sea salt

1 egg (beaten)

¼ cup coconut flour

1/3 cup shredded coconut (unsweetened)

- Combine all the dry ingredients in a bowl.

- Dip the chicken, first in egg and then dredge it in the coconut flour mixture.

- Place the chicken in a baking dish lined with parchment paper.

- Bake for 20-25 minutes in an oven preheated to 350 degrees Fahrenheit/ 175 degrees Celsius, flipping it once midway.

Per Serving: Calories: 393; Total Fat: 24.6g; Saturated Fat: 8.1g; Protein: 19.9g; Carbs: 22.9g; Fiber: 5g; Sugar: 1.5g;

Kale Chips

Serves: 2 / Preparation time: 10 minutes / Cooking time: 20 minute

Kale Chips are a great healthy evening snack. It's a welcoming way to have some of your greens when you don't feel like sitting down to a big plate of vegetables.

1 cup kale leaves (washed, torn, dried)   2 tablespoon olive oil

Sea salt to taste

- Toss all the ingredients in a baking sheet and spread them.

- Bake for 15 minutes in an oven preheated to 300 degrees Fahrenheit/ 149 degrees Celsius, tossing it once midway.

Per Serving: Calories: 137; Total Fat: 14g; Saturated Fat: 2g; Protein: 1g; Carbs: 3.5g; Fiber: 0.5g; Sugar: 0g;

Advice for Eating at Restaurants

At some point, you'll be eating at a restaurant. How do you stay faithful to your Paleo lifestyle? It actually isn't that difficult - it just requires a little research beforehand and the willingness to ask questions. Also, it's unlikely that you'll find some place that's 100% Paleo, but that's okay. Don't become stressed out if the best meal you can find has a certain cooking oil or if you can't find out where a piece of meat came from. If you stayed 100% Paleo all the time, you would never go out, and you'd probably never have a social life. It's okay to make a few compromises with your diet in order to maintain healthy relationships.

Here are some tips on staying as Paleo as possible:

BEFORE HEADING OUT

Research "Paleo restaurants" in your area

While you probably won't find a place devoted to Paleo food, you will find links to places that have Paleo-friendly options and reviews where diners have tagged the term "Paleo." Make a list of the places you find that have Paleo options. Other good terms include "locally-sourced meat," "grass-fed beef," "organic," and "farm to table."

Look up the menu

If you already have a specific eatery picked out, look up their menu to see what you can find that's Paleo-friendly. Places with lots of salad options, steaks, and fish are most likely to have something you can eat.

Call ahead

If you want to have certain questions answered about the type of cooking oil used or if substitutions are allowed, it's always good to call ahead. It gives the restaurant time to find the answers.

AT THE RESTAURANT

Be really nice to your server

The first thing to remember when you're eating Paleo and eating out is to be really nice to your server. They have a tough job, and you don't want to be on their bad side. Always be really polite when you ask questions about the menu. Leave a generous tip, too, if you have them going back and forth from your table to the kitchen.

No thanks to bread

A lot of restaurants have free bread or biscuits, so as soon as you get there, say, "If there's bread, we're going to pass on that." If they bring it to you before you can say anything, very politely say, "Oh, no thanks, we're not eating bread."

Avoid alcohol

When you're Paleo, you can't really drink alcohol. If you must drink, distilled liquors like vodka, rum, tequila, whisky, rum, brandy, or red wine are the best choices. Another alcohol option that's been getting more popular in the Paleo community is hard, dry cider. There's still sugar, but some brands (like Crispin Cider and Strongbow) have low-ish sugar counts. Ciders with high sugar include Angry Orchards and Woodchuck.

Start on gluten-free

Going gluten-free has become a lot more mainstream than going Paleo, so it's a good place to start when you're navigating a menu. There will still be dishes with dairy and other grains, but at least you've eliminated one big non-Paleo food group.

Look for buzzwords that scream "not Paleo"

There are certain food descriptions to watch for, because they always mean something is not Paleo. Terms include: deep-fried, battered, coated, breaded, and crispy. Anything with the words "fitter" and "dumpling," are also not Paleo, while meatballs and meatloafs might include breadcrumbs.

Substitute or eliminate side dishes + sauces

A lot of side dishes are going to include rice, pasta, or other foods that aren't Paleo-friendly, so simply substitute them for veggies or salads without dressing. Watch out for sauces, too, most are dairy-based, so ask for meals without it.

Order grass-fed burgers without buns

Ordering a burger without a bun is a really easy Paleo fix. You can also generally add or remove toppings. Ask about seasonings on the burger - a lot of places use spice blends that include artificial ingredients.

Order fajitas and tacos without the tortilla, rice, or beans

Another easy fix is to order Mexican food without the fajitas or tacos. If rice and beans are included, ask for those to be left off, too. You've basically made your own burrito bowl with meat and veggies.

At Asian places, eliminate rice and MSG

In general, Asian places can be tough, because they do rely on MSG, and often include ingredients like soy, peanuts, and added sugar. It'll likely take a lot of research to find a place you know for sure has Paleo options, though if you're really craving Asian food, it might be a chance to relax a bit about the Paleo restrictions. If you're at an Asian or Thai place, get a curry dish without rice or MSG. At sushi places, you can also find sushi rolls that don't have rice.

Find 1-2 places you really love, and stick to them

When you go Paleo, the best method is to find one or two restaurants that have a lot of Paleo options, and make them your go-to. You won't be eating out a lot on Paleo, so it's okay if the places are a bit pricier because of their quality.

BEST PALEO-FRIENDLY CHAIN RESTAURANTS

There are actually quite a few big chains that have Paleo-friendly options. Write down the ones in your area and stick it somewhere, so you always have a place to suggest to friends when they want to go out. At most places, you'll still have to double-check for yourself, and make sure they aren't using a forbidden cooking oil.

If you want to find more, check out the iPhone app PaleoGoGo. It's a database with other three hundred chains and gives you Paleo suggestions. Another app, HealthyOut, lets you set diet requirements, so you can set it for Paleo-friendly, ketogenic, and so on.

Here's a list of restaurants that have Paleo options:

- TGI Friday's
- Outback Steakhouse
- Applebee's
- Cracker Barrel
- Red Lobster
- Chili's
- Cheesecake Factory
- Jimmy John's
- Chipotle
- Moe's
- Shake Shack
- Panera
- Five Guys
- Subway
- In 'n Out

Celebrities Who Have Gone Paleo

The Paleo lifestyle is one of the most commonly-adopted eating lifestyles, especially among famous people. While celebrities do have professionals doing their makeup, hair, and wardrobe, you can't deny that they display the effects of a great diet. Many have followed in their favorite celebrity's footsteps, which no doubt contributed to the fact that "Paleo" was the most-searched diet in 2013. Nearly 5 years later, and the Paleo lifestyle is still extremely popular.

Here are seven celebrities that have gone Paleo and are loving it:

Jessica Biel

Jessica Biel has been Paleo for a while, and although she does have cheat days, she says she "just feels better" when she isn't eating grains and dairy. She's even opened a marketplace and restaurant called Au Fudge designed for kids and families, which is all-organic, and has vegan and Paleo options.

Anne Hathaway

Academy-award winning actress Anne Hathaway was vegan for many years, but while she was filming Interstellar, the physical work took its toll. She began eating animal proteins. After eating a single piece of fish, she was astonished at how much better she felt. These days, she eats a lot of salads and chicken.

Tim McGraw

In 2008, this iconic country singer realized his hard-drinking lifestyle was destroying his body. He gave up alcohol, went on the Paleo lifestyle, and began CrossFit training, which is very popular among the Paleo crowd. McGraw would eventually end up losing 40 pounds.

Megan Fox

Actress Megan Fox used to be a vegetarian, but felt pressured to put on more weight when people started to say she was "too skinny." Though her reason for adopting the Paleo lifestyle is unfortunate, Fox has enjoyed a lot of its benefits. After giving birth to her first son, Fox lost 23 pounds of baby weight very quickly. After just two months, she was back to her normal weight. She also credits the Paleo lifestyle with keeping her energized, which as a mother of now two kids, is very important.

Grant Hill

Athletes frequently go Paleo to increase their energy levels, and 7-time NBA All-Star Grant Hill is a prime example. He played professionally until 40, and says he felt better than he did at 30, thanks to the Paleo lifestyle. His mindset is simple: "If it was here a million years ago, then I tend to eat it."

Kobe Bryant

As basketball superstar Kobe Bryant reached the end of his career, he decided to shake up his nutrition to stay sharp. According to his trainer, Kobe's diet was packed with butter and grass-fed beef, while he cut down sugar and carbs significantly. He made the switch in 2012, and his game saw a marked improvement in the latter half of that season.

Matthew McConaughey

This charming Southern actor is known for his cut abs, so it makes sense that he pays close attention to his diet. He chose a Paleo lifestyle to stay healthy on a very busy schedule, and keep his body in shape for roles in films like "Magic Mike." He's been Paleo for quite a few years now, and keeps a garden at home with his wife.

30 Day Paleo Challenge Meal Plan

Living the Paleo lifestyle is not about slavishly counting calories or eating tiny portion sizes. As such, the following meal plan is not meant to be adhered to without deviation, nor is the intended idea that you eat only one serving of each recipe. If you want two servings, go for it! Also feel free to pair any meal with a side dish that appeals to you. You'll find lots of tasty options in the side dish chapter. Some days you might also feel like having a healthy Paleo snack to tide you over in between meals. And don't forget dessert! Flip to the dessert chapter and make something that sounds appealing to you after dinner whenever you feel like it.

Use this meal plan as a foundation that you can alter and build upon. One of the best features of Paleo is that it isn't a diet based on starving yourself, so go ahead and enjoy your meals!

DAY	BREAKFAST	LUNCH	DINNER
1	Breakfast Fritters, 50	Apple & Cherry Pork, 93	Chicken Stew, 76
2	Spiced Pumpkin Pancakes, 43	Chicken & Mushroom Toss, 74	Grilled Lamb Chops, 114
3	Hazelnut & Chocolate Cereal, 38	Coconut Buttered Haddock, 124	Beef Stew, 103
4	Pumpkin Spiced Puree, 42	Broccoli & Beef Mix, 104	Herbed Lamb Chops, 115
5	Milky Pancakes, 39	Citrus Chicken, 81	Salmon & Asparagus Bake, 127
6	Flax Meal Cereal, 46	Spicy Shrimp Soup, 60	Beef Meatloaf, 100
7	Scrambled Eggs with Fruit, 36	Spinach & Pear Salad, 67	Herbed Mahi Mahi Fillets, 123
8	Beef Muffins, 47	Ham Veggie Soup, 59	Hazelnut Smothered Cod Fillets, 121
9	Kale & Sausage Muffins, 40	Baked Nutty Chicken, 77	Lamb Rib Chops, 110
10	Green Mushroom Quiche, 48	Herb Flavored Veggies, 140	Turkey Stir Fry, 75
11	Nutty Banana Smoothie, 45	Roasted Shrimps with Veggies, 125	Turkey Meatloaf, 79
12	Choco-Zucchini Bread, 49	Lamb Shanks with Mushroom Gravy, 117	Lobster & Asparagus Mix, 126

13	Grapefruit & Carrot Smoothie, 37	Simple Onion & Tomato Salad, 68	Roasted Salmon with Leeks, 122
14	Beef Muffins, 47	Turkey & Veggie Toss, 78	Thai Style Ground Beef Curry, 98
15	Banana Pancakes, 41	Balsamic Pork Roast, 89	Sour & Spicy Beef, 102
16	Flax Meal Cereal, 46	Pork & Asparagus Stir Fry, 85	Beef in Marinara Sauce, 105
17	Pumpkin Spiced Puree, 42	Hot Chicken, 73	Spiced Lamb Leg, 109
18	Breakfast Fritters, 50	Salmon Crusted with Coconut, 129	Lamb Burgers, 113
19	Green Mushroom Quiche, 48	Barbacoa Lamb, 116	Lime & Garlic Pork Chops, 91
20	Spiced Pumpkin Pancakes, 43	Stuffed Mushrooms, 132	Lamb Stew, 112
21	Hazelnut & Chocolate Cereal, 38	Egg & Tuna Salad, 54	Spicy Chicken Chili, 72
22	Banana Pancakes, 41	Shrimps & Broccoli Stir Fry, 128	Salmon & Asparagus Bake, 127
23	Kale & Sausage Muffins, 40	Beef & Bell Pepper Stir Fry, 96	Pork Cutlets, 92
24	Choco-Zucchini Bread, 49	Lemon & Herb Shrimp Soup, 56	Taco Chicken Soup, 57
25	Nutty Banana Smoothie, 45	Pork & Cabbage Stir Fry, 84	Lamb Curry, 108
26	Scrambled Eggs with Fruit, 36	Beef Chili, 99	Herbed Pork Chops, 88
27	Milky Pancakes, 39	Mushroom Seafood Skillet, 120	Roasted Salmon with Leeks, 122
28	Grapefruit & Carrot Smoothie, 37	Roasted Shrimps with Veggies, 125	Lamb Shanks, 111
29	Flax Meal Cereal, 46	Salsa Chicken Wraps, 80	Beef Cutlets, 97
30	Breakfast Fritters, 50	Crispy Almond Crusted Pork, 90	Thai Pork Curry, 86

References and Resources

Boers, Inge, et al. "Favourable Effects of Consuming a Palaeolithic-Type Diet on Characteristics of the Metabolic Syndrome: a Randomized Controlled Pilot-Study."*Lipids in Health and Disease*, BioMed Central, 11 Oct. 2014, lipidworld.biomedcentral.com/articles/10.1186/1476-511X-13-160.

Daley, Cynthia A, et al. "A Review of Fatty Acid Profiles and Antioxidant Content in Grass-Fed and Grain-Fed Beef." *Nutrition Journal*, BioMed Central, 2010, www.ncbi.nlm.nih.gov/pmc/articles/PMC2846864/.

Fontes-Villalba, Maelán, et al. "A Healthy Diet with and without Cereal Grains and Dairy Products in Patients with Type 2 Diabetes: Study Protocol for a Random-Order Cross-over Pilot Study - Alimentation and Diabetes in Lanzarote -ADILAN." *Trials*, BioMed Central, 2 Jan. 2014, trialsjournal.biomedcentral.com/articles/10.1186/1745-6215-15-2.

Lustig, Robert H., et al. "Isocaloric Fructose Restriction and Metabolic Improvement in Children with Obesity and Metabolic Syndrome." *Obesity*, 26 Oct. 2015, onlinelibrary.wiley.com/doi/10.1002/oby.21371/abstract.

Mellberg, Caroline et al. "Long-Term Effects of a Palaeolithic-Type Diet in Obese Postmenopausal Women: A Two-Year Randomized Trial." *European journal of clinical nutrition* 68.3 (2014): 350–357. PMC. Web. 19 Oct. 2017.

"Paleo Leap | Paleo Tips and Recipes." Paleo Leap | *Paleo Recipes & Tips*, paleoleap.com/.

Sanfilippo, Diane. Practical Paleo: *a Customized Approach to Health and a Whole-Foods Lifestyle*. Victory Belt Publishing Inc., 2016.

Yang, PhD Quanhe. "Sugar Intake and Cardiovascular Diseases Mortality." *JAMA Internal Medicine*, American Medical Association, 1 Apr. 2014, jamanetwork.com/journals/jamainternalmedicine/fullarticle/1819573.

Recipe Index

Want MORE full length cookbooks for FREE?

We invite you to sign up for free review copies of future books!

Learn more and get brand new cookbooks for **free**:

http://club.hotbooks.org

Want MORE healthy recipes for FREE?

Double down on healthy living with a full week of fresh, healthy salad recipes. A new salad for every day of the week!

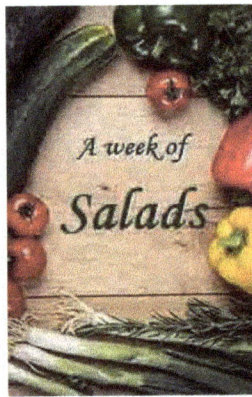

Grab this bonus recipe ebook *free* as our gift to you:

http://salad7.hotbooks.org